BRAVE
NEW
ARCTIC

Books in the *SCIENCE ESSENTIALS* series bring cutting-edge science to a general audience. The series provides the foundation for a better understanding of the scientific and technical advances changing our world. In each volume, a prominent scientist—chosen by an advisory board of National Academy of Sciences members—conveys in clear prose the fundamental knowledge underlying a rapidly evolving field of scientific endeavor.

BRAVE NEW ARCTIC

THE UNTOLD STORY
OF THE MELTING NORTH

MARK C. SERREZE

PRINCETON UNIVERSITY PRESS
PRINCETON AND OXFORD

Published by Princeton University Press,
41 William Street, Princeton, New Jersey 08540

In the United Kingdom: Princeton University Press,
6 Oxford Street, Woodstock, Oxfordshire OX20 1TR

press.princeton.edu

Jacket image: Baffin Island, Canada. Courtesy of Getty Images

Library of Congress Cataloging-in-Publication Data
Names: Serreze, Mark C., author.
Title: Brave new Arctic : the untold story of the
melting North / Mark C. Serreze.
Description: Princeton : Princeton University Press, [2018] |
Series: Science essentials | Includes bibliographical references and index.
Identifiers: LCCN 2017031687 | ISBN 9780691173993
(hardcover : alk. paper)
Subjects: LCSH: Climatic changes—Arctic regions.
| Arctic regions—Environmental conditions. | Arctic regions—Climate. |
Climatology—Arctic regions. |
Global warming. | Global environmental change.
Classification: LCC QC994.8 .S4754 2018 | DDC 577.27/609113—dc23
LC record available at https://lccn.loc.gov/2017031687

British Library Cataloging-in-Publication Data is available

This book has been composed in Baskerville 10 and Gin

Printed on acid-free paper. ∞

Printed in the United States of America

1 3 5 7 9 10 8 6 4 2

THIS BOOK IS DEDICATED TO
THE ARCTIC SCIENTISTS WHOM
I HAVE KNOWN AND WORKED
WITH FOR THESE MANY YEARS.

CONTENTS

PREFACE

As recently as the 1980s, the Arctic was, in many respects, the same Arctic that had enchanted humankind for centuries. But over the next decade, scientists from around the world began to notice changes. There were hints that the floating sea-ice cover at summer's end was receding, accompanied by shifts in ocean circulation. Air temperatures over some parts of the Arctic were distinctly rising, although other areas were cooling, attended by puzzling changes in weather patterns. Permafrost—the Arctic's perennially frozen ground—showed signs of warming. Although it had long been recognized that the human imprint on climate would likely appear first in the Arctic, much of what was happening had the look of a natural climate cycle. Still, the changes kept coming. Through a largely self-organizing process, scientists from diverse disciplines and from around the world began to find the answers. There were remarkable discoveries, periods of confusion, and controversy. Through their efforts, by the second decade of the 21st century, the picture had cleared. We were well on our way toward a warmer and profoundly different North, essentially free of summer sea ice, with effects on climate and human systems potentially spanning the globe. This book tells the story of the melting North. It draws strongly from my own perspective as a climate scientist who saw it all happen and from those whom I have known and worked with for these many years.

LIST OF ACRONYMS

ACIA   Arctic Climate Impact Assessment
AO   Arctic Oscillation
AON   Arctic Observing Network
ARCSS   Arctic Climate System Study
ASTER   Advanced Spaceborne Thermal Emission and Reflection Radiometer
ATLAS   Arctic Transitions in the Land-Atmosphere System
AVHRR   Advanced Very High Resolution Radiometer
CHAMP   Community-wide Hydrologic Analysis and Monitoring Program
DMSP   Defense Meteorological Satellite Program
EOF   Empirical Orthogonal Function
FWI   Freshwater Integration
GRACE   Gravity Recovery and Climate Experiment
IABP   International Arctic Buoy Programme
IARC   International Arctic Research Center
ICESat   Ice, Cloud, and Land Elevation Satellite
IPCC   Intergovernmental Panel on Climate Change
IPY   International Polar Year
LDGO   Lamont-Doherty Geological Observatory
MODIS   Moderate Resolution Imaging Spectroradiometer
MOSAiC   Multidisciplinary Drifting Observatory for the Study of Arctic Climate
NAM   Northern Annular Mode
NAO   North Atlantic Oscillation
NASA   National Aeronautics and Space Administration
NCAR   National Center for Atmospheric Research
NCEAS   National Center for Ecological Analysis
NDVI   Normalized Difference Vegetation Index
NOAA   National Oceanic and Atmospheric Administration

LIST OF ACRONYMS

NPEO   North Pole Environmental Observatory
NSF   National Science Foundation
NSIDC   National Snow and Ice Data Center
ONR   Office of Naval Research
PIOMAS   Pan-Arctic Ice-Ocean Modeling and Assimilation
    System
SEARCH   Study of Environmental Arctic Change
SHEBA   Surface Heat Balance of the Arctic Ocean

BRAVE
NEW
ARCTIC

1

BEGINNINGS

Turning points in life are seldom recognized until they have already passed. In my case, that turning point was in 1981. After a series of aimless years, I finally landed on a track toward a bachelor's degree from the University of Massachusetts Amherst in physical geography. I'd started out in 1978 as an astronomy and physics major, but for a number of reasons, none of which bear especially close scrutiny, I decided to go in a different direction. On the plus side, it was clear that a bachelor's in geography was better than no degree at all. On the minus side, I hadn't yet learned enough hard science to be employable, only enough to be irritating to my friends.

Lucky for me, the decision panned out. I ended up being in the right place at the right time to seize an opportunity and see part of the world where, at the time, few had ventured. Six months later, I found myself in a ski-equipped Twin Otter headed to northeastern Ellesmere Island in the Canadian High Arctic to begin a detailed study of two little ice caps. I became

1

enchanted with the North and decided to become an Arctic climatologist. By 2016, those ice caps had almost completely melted away, victims of the Arctic meltdown. I could never have imagined this at the time. I could not have known that in becoming a climate scientist, I was to earn a front-row seat to observe how, in fits and starts, it first began to be noticed that the Arctic was changing. Nor could I have known that I'd also become part of the growing cadre of scientists who first struggled with conflicting evidence to try and make sense of what was happening, then finally had no recourse but to yield to the conclusion that a radical transformation was underway. I could not have foreseen that Arctic climate research, once the domain of a small community of scientists with love for snow and ice, would become a centerpiece in the quest to understand the impacts of global climate change that would involve collaboration between thousands of scientists from around the world.

CHARTING A COURSE

It was a rainy afternoon when I learned that Dr. Raymond Bradley, an associate professor at the Department of Geology and Geography, was teaching upper-division courses in both climatology and paleoclimatology—climates of the past.[1] This sounded like interesting stuff, so I signed up for both.

Since elementary school, I had been aware that the earth's climate had varied in the past, but until taking Ray's courses I had no real idea how these variations related to things like periodic changes in earth-orbital configuration, atmospheric greenhouse gas composition, volcanic eruptions, solar variability, and climate feedbacks. Ray drew in part from his own research, which focused on the past and present-day climate of the Arctic. Ray wrote his first research paper in 1972 while still a graduate student.[2] He found that a global warming trend starting in the 1880s, particularly notable during the winter season and in the Arctic, changed to a cooling trend in the 1940s. He later documented a rather abrupt further cooling in the Canadian High Arctic starting right around 1963/1964, which he suspected might relate to a massive injection of dust into the upper atmosphere from the 1963/1964 eruption of Mount Agung, a rather ill-tempered and still active volcano located in Bali, Indonesia.[3] The cooling noted by Ray and others turned out to be a temporary thing, but for a time it helped to foster speculation, greatly overstated by the media, that the planet might be entering a long-term cooling phase. Reflecting my fondness for big snowstorms and seeing commerce grind to a halt, I found the idea of a cooling planet quite appealing. While part of the climate class also covered the already quickly growing counterpoint that because of the observed rise in carbon dioxide levels in the atmosphere, as measured at the Mauna Loa Observatory, the planet should start to warm up, and most strongly in the polar regions, deep down I was hoping for an ice age.

I was friendly with Mike Moughan, a fellow a few years older than me who was one of Ray's graduate students. Making full use of the university's CDC Cyber Systems mainframe computer, Mike was processing temperature and precipitation data from weather stations across the Canadian Arctic (with enchanting names like "Resolute Bay," "Alert," and "Eureka") to better understand variability and recent trends in the region's climate. His work doing real climate research seemed so cool, and he looked so scientific walking down the hall of the Morrill Science Center with computer printouts or toward the Computing Center carrying a 9-track magnetic tape of valuable data.

I wanted to be part of it. The opportunity came when Mike decided that he was not up for graduate school. This left Ray in a lurch. Upon Mike's suggestion, Ray agreed to take me on as an hourly student, at a seemingly princely wage of five dollars per hour, to finish the work that Mike had started. Mike showed me how to log onto the CDC Cyber Systems mainframe, and how to edit the SPSS routines that he had been using. After climbing a steep learning curve, I became competent enough to supply Ray with data plots. Now I was the cool dude walking down the hall and to and from the Computing Center.

In early 1982, Ray inquired about my future plans and said that if I was up for it, he needed a field assistant for the upcoming summer's work in the Arctic. I enthusiastically volunteered. He also emphasized that I ought to apply to graduate school and take Mike's place. I applied.

Ray's project was to reconstruct the past glacial history of the Queen Elizabeth Islands, which is a part of the Canadian Arctic Archipelago. At the time, this area was a part of the Northwest Territories; it is now part of Nunavut. The project involved recovering and analyzing sediment cores from Arctic lakes, including a series of small freshwater bodies called the Beaufort Lakes, near the northeastern coast of Ellesmere Island. Ray had been coordinating his research with Dr. John England from the University of Edmonton.

Via a well-written proposal, Ray convinced the U.S. National Science Foundation (NSF) to support a modest additional project on and around a pair of nearby small, stagnant ice caps at about 1000 m elevation on the Hazen Plateau (fig. 1). The NSF, as I quickly learned, is the key federal agency supporting fundamental research and education in the non-medical fields of science and engineering; its counterpart in medical fields is the National Institutes of Health.

The objective of this side project was to shed light on an idea advanced in 1975 by Jack Ives of the University of Colorado Boulder regarding how the great continental ice sheets of the Pleistocene might have formed.[4] It had long been known that the past 2 million years or so had seen a series of major ice ages, separated by warm

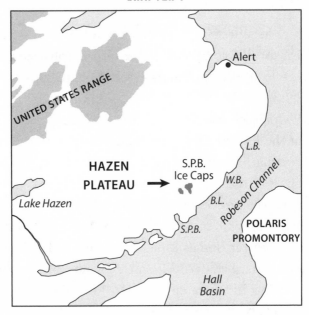

FIGURE 1: Northeastern Ellesmere Island and the location of the St. Patrick Bay ice caps on the Hazen Plateau, near St. Patrick Bay (S.P.B.) and the Beaufort Lakes (B.L.). Source: Bradley, R., & Serreze, M.(1987), "Topoclimatic Studies of a High Arctic Plateau Ice Cap." *Journal of Glaciology*, 33(114), 149-158.

interglacials, like the one we live in today. Ives's thinking was that the past great ice sheets of North America, the most recent being the Laurentide Ice Sheet, at its biggest about 25,000 years ago, initially formed through the accumulation of snow on the extensive Labrador-Ungava plateau of Canada. If the climate cooled for some reason, then the snow line would drop below the altitude of much of the plateau surface. Temperatures tend to decrease the higher one goes in altitude, and above a certain altitude, it is cold enough that the snow that falls during winter survives the summer melt season. This elevation determines the snow line.

The drop in the elevation of the snow line below the level of the plateau surface would raise the reflectivity of the surface (that is, its albedo), reducing how much of the sun's energy is absorbed, further cooling the climate over the plateau, fostering the survival of even more high-albedo snow the next summer, and so on. The snow would eventually compress into ice, forming glaciers that would then coalesce, eventually growing to an ice sheet. Because the initial snow cover would quickly expand via this albedo feedback mechanism, the process was dubbed, with considerable exaggeration, instantaneous glacierization. As early as 1875, James Croll, in his book *Climate and Time in Their Geological Relations: A Theory of Secular Change of the Earth's Climate*, had recognized albedo feedback as an important climate process. He saw that the whole thing could work in reverse as well—warm conditions lead to less snow and ice, lowering the albedo and favoring more warming.

The roughly cyclical timing of past ice ages and interglacials implied a climate force that was itself cyclic. Using ocean core records, in 1976, James Hays, John Imbrie and Nick Shackleton presented convincing evidence that the major ice ages and interglacials of the Pleistocene had been "paced" by variations in earth orbital geometry called Milankovitch cycles.[5] Named after the Serbian geophysicist and astronomer Milutin Milankovitch, these cycles refer to variations in the earth's orbital eccentricity (departure from circular), its obliquity (tilt), and the timing of the equinoxes (precession) that affect how much solar energy reaches the top of the atmosphere at different latitudes and at different times of the year. Although astronomical theories to explain climate change had been around since the 19th century, they had not been verified by observation. The view of Milankovitch cycles as a pacemaker also recognized that orbital conditions favoring ice sheet onset (in particular, cool summers over the higher latitudes of the Northern Hemisphere) would then kick in various climate feedbacks to hasten the cooling, albedo feedback being but one of them. It is now known that carbon feedback is a biggie—as it cools, carbon dioxide comes out of the atmosphere and is stored in the oceans, and further cools the climate.

While Milankovitch effects had nothing to do with the temporary change toward Arctic cooling that Ray discussed in his 1972 paper, the cooling, through its potential link with albedo feedback, was one of the key science themes driving the ice cap study. "How misguided that looks nowadays," recalls Ray regarding the cooling phase, "though at the time it was pretty accurate—cooler and wetter winters on Baffin Island, and colder summers, so upland snow cover was indeed expanding."

The strategy of the NSF-sponsored ice cap study was to set up a weather station on top of the bigger of the two ice caps to measure air temperature, solar energy fluxes, albedo, and other variables. We would compare these to other measurements collected at stations set up at different distances beyond the edge of the ice cap at a similar elevation. Looking at the differences would tell us how the ice cap was affecting the local climate and how far the effects extended beyond its margins. It amounted to a local evaluation of some of the ideas encapsulated in instantaneous glacierization. In the spring of 1982, I invested a lot of time testing the instruments and the state-of-the art data loggers (called Microloggers) from the Campbell Scientific company.

OFF TO THE ARCTIC

We left for Ellesmere Island in May 1982. Beforehand, we'd shipped the major equipment to Resolute Bay in the care of the able government-run Polar Continental Shelf Program that handles logistics in the Canadian far north, directed for many years by Canadian scientist George D. Hobson.[6] I left first, accompanied by Mike Retelle, another of Ray's graduate students, and his assistant, Dick Friend, who would be staying with Mike at Beaufort Lakes for the coring work. We flew from Bradley Field outside of Hartford, Connecticut, to Montreal, and boarded a lumbering 737-200 to Edmonton, Alberta, operated by Pacific Western ("Piggly-Wiggly") airlines. We spent two boozy nights in Edmonton with one of John England's graduate students; we spent days visiting various stores, getting last-minute supplies together. Ray flew into Edmonton a day or so later. He informed me that although I had forgotten to ship the portable generator, a rather severe oversight, I had been accepted into the graduate school.

The next morning, we boarded the twice-weekly Piggly-Wiggly flight to Resolute Bay with a stop in Yellowknife. The specially equipped 737-200 C landed at Resolute Bay in a cloud of dust and gravel. The plan was to be ensconced for a few days at the ugly yet functional Polar Continental Shelf Program building, then head to Beaufort Lakes and the ice caps in a ski-equipped Twin Otter. Aircraft time is expensive. To save money, we would coordinate logistics with John England's group from Edmonton; they would be doing work on and around Polaris Promontory, Greenland, just across the Robeson Channel, which is the narrow ocean channel separating northeastern Ellesmere Island from northwestern Greenland (fig. 1).

Because of bad weather, a few days at Resolute Bay turned into almost a week. We spent the days eating, reading, eating, reading, eating, and moseying down to the weather station to look at the forecasts. We spent evenings at the Resolute Bay Bar. The bar, patronized by the local Inuit, base personnel, civilian and military pilots (Royal Canadian Air Force), and whoever else was around, was something right out of a Robert Service poem—dingy, dark, smoky, raucous, sexist, and not entirely safe.

Finally, the weather started looking better, and we headed out. Twin Otters fly low and slow, and we were looking at about three hours to our destination. However, the weather shut down again, and we diverted to Eureka, lying to the west. The routine over the next few days was pretty much the same as that at Resolute Bay, including nightly visits to the somewhat more upscale RCAF bar. A rule at the RCAF bar was that anyone caught wearing a hat, as I unwittingly did when first entering, was required to buy a round of drinks for everyone. Only by repeatedly pleading ignorance as an American civilian did I escape the sentence, which was very fortunate, given my limited bankroll.

The weather settled again, and Ray and I flew out ahead of the rest of the team. The Twin Otter landed on the ice cover of the largest of the Beaufort Lakes (a pond, really), and the gear was unloaded in short order. The plane headed back to Eureka, picked up Mike and Dick and the rest of the gear, and safely landed on the ice-covered lake a second time. We spent the next two days setting up camp for the Beaufort Lakes party and organizing. The weather continued to hold. The same Twin Otter returned from Eureka, picked up Ray and me with our gear, and made the short hop to the Hazen Plateau and the larger of the two ice caps. It was a rare cloudless, windless day. The temperature was probably 5°F or so, and the fresh snow on the plateau sparkled. The pilot dropped us off with our food, gear, white gas for the stove, regular gas for the generator, and a two-way radio, and then roared off.

It took about a week to set up camp and the weather stations. We had a large aluminum-framed igloo-style tent for sitting and cooking, and a couple of smaller tents for sleeping (fig. 2). A Coleman stove in the igloo tent served for cooking and melting snow for drinking water, and as a source of heat. Potential death by carbon monoxide poisoning never entered our minds. The sleeping tents were unheated, but we had warm sleeping bags.

Once everything was up and running, attention turned to a detailed survey of snow conditions on the ice cap. The Hazen Plateau, like almost all of the Canadian Arctic Archipelago, is a very dry environment, classified as a polar desert. The average total annual precipitation is on the order of only 20 centimeters (less than 8 inches), but because it is such a cold environment, evaporation is also quite low. Hence, in summer, it can be an oddly damp desert. Snow depths on the ice cap were typically in the range of 30–50 cm, representing almost all of the precipitation that had fallen since the end of the previous summer. Every so often, we measured the water equivalent of the snowpack. This required sticking a snow-coring tube through the snowpack to its base, recording the depth of the snow, extracting the coring tube along with its sample of snow, dumping the snow sample in a plastic bag, and then weighing the bag of snow. Knowing the snow depth and cross-sectional area of the tube, we could determine the snow volume. By measuring the weight (more properly, the mass), we could also get the snow density and water equivalent of the snow—that is, how much actual water is contained in

FIGURE 2. Camp at the edge of the ice cap, early June 1982, with a full snow cover. Courtesy of the author.

the snow. These numbers told us how much accumulation there had been on the ice cap through the previous autumn and winter.

The next step was to insert a series of aluminum stakes into the ice, using a hand-powered ice drill. By measuring the distance from the ice surface to the top of the stakes during spring (before melt starts) and then again at the end of summer, and combining this with information from the snow surveys, we could determine how much melt had occurred during the summer. The difference between the autumn/winter mass gain and the summer snow and ice loss represented the annual mass balance of the ice cap. A positive mass balance meant a growing ice cap (because of more autumn/winter accumulation than summer melt); a negative balance (more summer melt than autumn/winter accumulation) meant a shrinking ice cap.

Back in 1972, Canadian scientists Harold Serson and J. A. Morrison, despite foul weather, managed to insert eight aluminum stakes in transect partway across the ice cap—the start of a stake network. Late in that same summer, by which time the melt season had pretty much ended, the ice cap was again visited by Geoffrey Hattersley-Smith and A. Davidson. They found the ice cap and surrounding plateau to be completely snow-covered, pointing to a positive mass balance for that year.[7] This was very different from the situation in 1959, when high-altitude air photographs showed the ice caps to be free of snow, with exposed dirt layers standing out sharply against the dark tundra, pointing to a negative mass balance for that year.

Once our initial survey was done, things settled into a satisfying routine, with odd hours because of the 24 hours of sunlight. With the arrival of summer, snow started to melt off the surrounding tundra, and later off the ice cap. The data from the weather stations showed that the ice cap was having a strong impact on the local climate, which meant that I would have something to write about for my master's thesis. We conducted further surveys and explored the surrounding area, including the smaller of the two ice caps.

I was totally into it and took great pride in my measurements and in providing precise weather reports to the Polar Continental Shelf Project base, with impeccable radio etiquette. Passion bordered on the disturbing, such as one day when, despite a complete whiteout, with visibility of perhaps 100 feet and, of course, no GPS, Ray and I tried to hike out to download data from the Microloggers. After half an hour, we came upon footprints in the snow. We were shocked. Who else could possibly be out here? Russian spies? But the footprints were ours! We had walked in a circle.

In the evenings under the midnight sun, nursing valuable rations of scotch, Ray related stories of the age of early Arctic exploration. There were stories of scientific triumph, such as Fridtjof Nansen's idea of freezing his stout little ship, the *Fram*, into Arctic Ocean pack ice off the coast of the New Siberian Islands, letting it drift with the currents to determine the basic circulation of the ocean. But many were of tragedy. Ray, being British, was enamored with the disappearance of Sir John Franklin's ships, the *Erebus* and the *Terror*, which set out in 1845 to conquer the fabled Northwest Passage—the shortcut between the Atlantic and Pacific Oceans through the channels of the Canadian Arctic Archipelago—but failed to emerge in the Pacific, and the later discovery of relics from the expedition and gravesites on Beechey Island, where it appears that the crew of originally 129 men over-wintered in 1845/1846. There was evidence that as the end neared, some of the men resorted to cannibalism, shocking Victorian England. During the 1903–1906 voyage of the *Gjøa*, Roald Amundsen finally conquered the Northwest Passage with six companions, but it took them two and a half years to navigate the ice-choked channels. Ray was adamant in his belief that Robert Peary never made it to the North Pole, but Frederick Cook's claim of being the first to the pole a year earlier than Peary, in April 1908, seemed at least equally dubious.

Most fascinating to me as an aspiring scientist (or "cub scientist," as Ray would say), was the story of Lt. Adolphus Greely's Lady Franklin Bay expedition to Discovery Harbor, Ellesmere Island, as part of the First International Polar Year (IPY). The IPY, which took place in 1882/1883, was the first major international effort to collect data to better understand the Arctic environment. Twelve scientific stations were established, including Fort Conger at Discovery Harbor, just down the coast from St. Patrick Bay. While Greely is known to have explored much of the area around Fort Conger, he never mentioned the ice caps. Perhaps the extensive snow cover over the plateau during the colder conditions of the late 19th century—the tail end of the Little Ice Age—masked their presence. The Fort Conger site had actually first been used as a wintering site by the crew of the *HMS Discovery* during the British Arctic Expedition of 1875, led by George Nares. The expedition was an attempt to reach the North Pole via Smith Sound. Although it failed to reach the pole, the *HMS Discovery*, along with a second ship, the *HMS Alert*,

THE GREELY EXPEDITION, 1881.

H. Biederback, rescued.   Sergt. Cross, died Jan. 8, 1884.   Sergt. Linn, died Apr. 6, 1884,   Chas. Henry, died June 6, 1884.   Sergt. Ralston, died May 23, 1884.   Dr. Pavy, died June 6, 1884.   Sergt. Gardner, died June 12, 1884.

Wm. Whistler, died May 24, 1884,   S. Bander, died June 16, 1884.   Sergt. Fredericks, rescued.   W. A. Ellis, died May 19, 1884.   Sergt. Long, rescued.   Corp. Salor, died June 3, 1884.   Sergt. Ellison, died July 6, 1884.

Private Connell, rescued.   Sergt. Brainard, rescued.   Lieut. Kislingbury, died June 1, 1884.   Lieut. Greely, rescued.   Lieut. Lockwood, died Apr. 9, 1884.   Sergt. Israel, died May 27, 1884.   Sergt. Jewell, died Apr. 13, 1884.   Sergt. Rice, died Apr. 9, 1884.

FIGURE 3. Members of the Lady Franklin Bay expedition. Greely is sitting in the bottom row, fourth from the left. Courtesy of National Archives, photo no. 200-LFB-134.

explored large parts of the coasts of both Greenland and Ellesmere Island.

The order given to Greely was that if relief ships failed to reach him, he was to retreat with his men southward down the coast. While his party would have been able to make it through winter quite well at Fort Conger, given the plentiful game in the area, Greely carried out his orders. They retreated, and winter overtook the party. Of the original expedition members (fig. 3), only seven, including Greely, lived to meet the relief ship in June 1884. Some of the meteorological records nevertheless survived.

GAINING EXPERIENCE

Our plan was that partway through the field season, Ray and I would fly to Polaris Promontory, Greenland. There, Ray would team up with John England. Meanwhile, I would fly to Beaufort Lakes with a fellow named Chris, and then hike back to the ice caps. This plan got delayed by an impressive blizzard that kept us holed up. On July 4, in good weather, we hiked down from the ice caps about 10 miles to a raised proglacial delta near St. Patrick Bay where, by previous agreement established by radio, a Twin Otter would be waiting. The plane arrived as planned, and we clambered on board, flew to Polaris Promontory, and dropped off Ray. Chris and I returned to Beaufort Lakes, spent the night with Mike Retelle and Dick Friend, and then hiked back up to the ice caps. Mike and Dick went part of the way up with us. Along the way, we were attacked by a group of four ill-tempered musk oxen. A shotgun blast over their heads, meant to dissuade them, instead provoked ire, and we were obliged to retreat to the safety of a small, steep hill. Chris and I eventually made it back to the ice cap in good order and took advantage of the half bottle of scotch reserved for the triumphant arrival.

We expanded the stake network and conducted surveys of the snow and ice conditions every week. It seemed that it was to be a negative mass balance year. During our expedition, all of the snow melted off, leaving bare ice with a cryoconite surface. Cryoconite is a powdery, windblown dust, comprised variably of small rock particles, soot, and microbes. It is dark and hence has a much lower albedo than the ice. It tends to clump to create small holes, which in the case of the ice cap, were typically a few centimeters deep and 5–10 centimeters in diameter. I had never seen anything like it. The tents, initially set up right near the edge of the ice cap, became wet and miserable. Eventually, we moved them to a better location. The food selection slowly narrowed to a choice between cans of tuna fish, Kraft macaroni and cheese, and canned French-style green beans. Sanitary conditions were deplorable, but temperatures generally near the freezing point helped in this regard. By the time we left the ice cap at the end of July, with autumn already setting in on the Hazen Plateau, I knew every nook, cranny, elevation change, and mood of the ice cap. I felt a sense of responsibility and ownership.

A helicopter took us down to Fort Conger, where we joined Mike Retelle, Dick Friend, and John England. Waiting for the Twin Otter to land at the short airstrip and take us back to Resolute, we explored the ruins of the fort and the surrounding area. After Greely left on his ill-fated retreat down the coast, Fort Conger stood idle until Robert Peary stopped there during his unsuccessful 1899 expedition to reach the North Pole. Peary again visited Fort Conger in 1905 and 1908. He tore down the original three-room fort and used the wood to construct several smaller buildings better suited to the environment. In later years, other expeditions used the site. A fascinating sight during our several-day stay was the stack of numerous bleached skulls of musk oxen shot for food back in the days of Greely and Peary.

The year rolled by and in April 1983, I gained additional experience as an assistant to the renowned Canadian glaciologist Roy (Fritz) Koerner, doing mass balance measurements on the Devon Ice Cap, on central Ellesmere Island, and on the Meighen Island Ice Cap.[8] A case of frostbite on my nose and face while at the top of the Devon Ice Cap instilled wisdom regarding always having the proper gear. I also briefly visited the Ward Hunt Ice Shelf on the northern end of Ellesmere Island, the likely source of a number of tabular (flat-topped, like a table) icebergs that had been discovered floating around the Arctic Ocean, some of which found use as platforms for scientific observations, such as Hobson's Choice and T3 (also known as Fletcher's Ice Island, named after its discoverer, U.S. Air Force Colonel Joe Fletcher). Ice shelves form when glacial ice flows out into the ocean, forming a flat floating sheet. Tabular icebergs occasionally break off from their seaward edge. While Ward Hunt is tiny compared to the ice shelves surrounding the Antarctic continent (which sometimes produce tabular icebergs the size of Delaware, such as iceberg A68, which calved from the Larsen C ice shelf in summer 2017), it was the largest one in the Arctic. Very little of it is left today.

Ray and I got back to the ice caps in late May 1983, with a new field assistant (yet another Mike) in tow. We set up camp and the weather stations, and awaited the melt season. The data showed that the mass balance of the ice caps in 1982 had indeed been negative, with all the snow melting off and some of the bare ice melting away as well. So maybe the Arctic was warming up, as expected. But by sharp contrast, just like Hattersley-Smith and Serson had experienced in 1972, the snow never melted away in 1983. I felt vindicated, and we jokingly referred to 1983 as the Year of Instantaneous Glacierization. The stake network was further expanded in anticipation of future visits to assess the mass balance (fig. 4), and we collected high-quality weather data. For my master's thesis, I planned to look at the effects of the ice cap on the local climate for two contrasting years. Ray left partway through the summer, and Mike and I finished a productive field season. We packed everything up in late July and, as before, flew by helicopter down to Fort Conger, to await the arrival of the Twin Otter the next day to take us back to Resolute. Again, our food supply was very low. On the last evening, we dined on two cans of French-style green beans, and an Arctic hare illegally and rather messily dispatched with the shotgun.

I never returned to the ice caps. I wrote my master's thesis; Ray insisted that it should weigh at least a kilogram, and at over 200 pages on heavy thesis bond paper, it passed muster. We eventually wrote a couple of papers on the mass balance and the energy balance of the ice caps, and then I moved on to eventually earn a PhD

FIGURE 4. Field assistant M. Palecki drilling into the ice during expansion of the stake network in 1983. Courtesy of the author.

at the University of Colorado in geography in 1989 and embark on a career in Arctic climate science.

GOING, GOING, GONE

More than 30 years have passed, and the Arctic I visited back in the early 1980s is fading in the rearview mirror. While I had not forgotten about "my" little ice caps, there was so much going on in the Arctic that I hadn't really kept up with how they were doing. In the spring of 2016, on a whim, I browsed through online satellite images from the NASA MODIS instrument (Moderate Resolution Imaging Spectroradiometer) to check up on them. I couldn't locate the ice caps, so I walked down the south hallway of the National Snow and Ice Data Center (NSIDC) to the office of my colleague Bruce Raup. Bruce was involved in an international project to map the world's glaciers and ice caps using satellite data at higher spatial resolution (15 m) from a NASA instrument called ASTER (Advanced Spaceborne Thermal Emission and Reflection Radiometer). I gave him the coordinates, and we looked for a while to find summertime clear-sky images for recent years when the bright ice caps ought to be standing out prominently against the plateau surface. We finally found them, and I could not believe what I was looking at. They had nearly disappeared.

Back in 1959, when aerial photographs were taken, the larger ice cap had an area of 7.48 square kilometers and the smaller one about 2.93 square kilometers. In August 2001, University of Massachusetts scientists Carsten Braun and Doug Hardy, two of Ray Bradley's later generation of graduate students, had returned and conducted a detailed survey of the ice caps. They found some of the aluminum stakes that had been so laboriously drilled into the ice, but all of them had melted out and were lying on the ground. They measured the perimeter of both ice caps using portable GPS. By 2001, the larger and smaller ice cap had already shrunk in area, respectively, to 62% and 59% of their 1959 areas.[9] This I had known. However, the NASA ASTER satellite data showed that as of July 2016, both of the ice caps

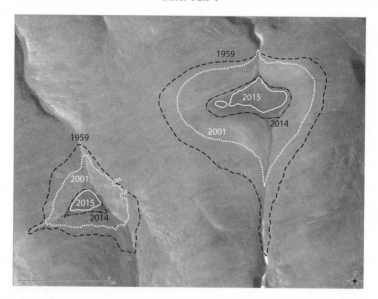

FIGURE 5. The St. Patrick Bay ice caps as seen on August 4, 2015, from ASTER imagery, along with the margins of the ice caps for August 2014 (also from ASTER), 2001 (from GPS surveys by C. Braun and D. Hardy), and 1959 (from aerial photography). Courtesy of the author.

covered only 5% of the areas that they did back in 1959 (fig. 5)! They are just ice patches now—the term "ice cap" is being kind. Quite remarkable is how they noticeably shrank even between 2014 and 2015. This appears to have been in direct response to the especially warm summer of 2015 over northern Ellesmere Island. What remains of the ice caps will likely vanish in only a few years.

From the available evidence, the ice caps probably formed back in the Little Ice Age (about 1650–1850), and at maximum extent they were several times larger than observed in 1959. They have been in an overall state of decline ever since, interrupted by periods of growth. They may have eventually disappeared without our help, but that is a moot point. I've been to a lot of interesting places since 1983, but since looking at the ASTER data, a day has rarely gone by that I haven't thought about my early adventures on those ice caps. As they melt away, part of me is also dying.

UNCHARTED WATERS

The climate records are clear. Surface temperatures over the Arctic as a whole are rising twice as fast compared to the rest of the globe. The Arctic is quickly losing its summer sea-ice cover, and wintertime losses are also starting to become prominent. Permafrost is warming and in some areas is thawing. Arctic glaciers and ice caps and the Greenland ice sheet are all losing mass, contributing to sea-level rise. The Ward Hint ice shelf that I visited back in 1983 is almost gone. Snow arrives later in fall, and melts earlier in spring than it used to. The character of precipitation is changing, with rain-on-snow events leading to massive reindeer mortality episodes.[10] Arctic ecosystems are shifting, and recent years have seen unprecedented heat waves over the Arctic Ocean during autumn and winter. The forces of change seem to be unstoppable. Looking forward, well within this century, perhaps only 20 or 30 years from now, the Arctic Ocean will be essentially free of its floating sea-ice cover in late summer. Sea ice will hence be but a seasonal feature. The two little ice caps near St. Patrick Bay that set me on a path to a career in Arctic climate science will be long gone.

For many people, the Arctic seems like such a far-away place, and given the more immediate challenges that they may be facing, such as putting food on the table and keeping a roof over their head, the changes unfolding in the North may seem unimportant. This is understandable. But there are also those who, for various reasons, choose to ignore what is happening. Then there are those who, largely out of self-interest, are in denial and maintain that it's all part of some natural climate cycle. Some even take the astonishing stand that the scientists are somehow making it all up or blowing it out of proportion. This is foolish.

It can be unpleasant to wake up to reality, but the Arctic is sounding alarm bells that cannot be ignored, and there is no snooze button. What the Arctic is telling us is that climate change is not some vague threat somewhere out in the future that may not even turn out be real, but is rather already here and here in a big way. The meltdown of the North is a clear demonstration that humanity has come to the point where we are geo-engineering our own planet—we have entered the Anthropocene. As we'll learn in coming chapters of this book, even as recently as the early 1990s, the Arctic largely seemed like the Arctic of old, and it took at least a decade for the science community—including myself—to fully come around to the inescapable conclusion that the region was being transformed. And since the dawn of the 21st century, the changes have been coming ever faster and are ever more troubling. The magnitude and scope of the changes that are unfolding have shaken the science community to its roots. We have entered uncharted waters.

If the theme of this book could be summed up in a single word it would be *complexity*. The complexity of the Arctic system is the very reason there are still so many unknowns out there regarding not just the Arctic's future, but impacts beyond its boundaries. Arctic change is certainly affected by what is happening in lower latitudes, but does it go the other way? Will Arctic amplification—the outsized warming of the Arctic compared to the rest of the globe, affect weather patterns in lower latitudes? Has this already happened? Will thawing permafrost lead to a large release of carbon back to the atmosphere, exacerbating the warming not just in the Arctic but for the planet as a whole, and if so, when will this start and how strong will the effect be? Melt of the Greenland ice sheet and of Arctic ice caps and glaciers will certainly continue to contribute to rising global sea levels, but by how much? One pretty sure thing is that as the Arctic continues to become more accessible, it will be a busier place, with less sea ice opening up shipping routes and making rich stores of oil and natural gas under the Arctic seafloor more accessible. Conflicts may arise.

CHAPTER 1

How did we get to this point? To answer this question, it is necessary to first take a close look at the Arctic of today, then step back again in time, to when, in fits and starts, it first began to be noticed that the Arctic was stirring.

2

IT'S NOT WHAT IT USED TO BE

Viewed in a global context, the imminent death of the St. Patrick Bay ice caps is inconsequential. But their fate is symptomatic of a remarkable transformation of the Arctic environment over the past several decades encompassing the land, the oceans, and the atmosphere. The Arctic has become a hotbed of scientific research and a constant source of media attention. The Arctic's soul is its cryosphere—its ice in all of its forms, which includes the floating sea-ice cover of the Arctic Ocean, snow, the Greenland ice sheet, ice caps and glaciers (collectively, glacial ice), ice on lakes and rivers, and permafrost. Whatever lives in the Arctic—be it flora or fauna—has adapted to coexist with the cryosphere. There is hence an important number—the temperature relative to the melting point of water, which is 32°F or 0°C. So much of what has happened in the Arctic reflects an upward shift in temperature with respect to that critical number. The melt season is becoming longer and more intense, and winter cold is fading. The loss of snow and ice then affects the warming because of climate feedbacks. This is part of the phenomenon of Arctic amplification—the faster warming of the Arctic compared to the rest of the planet. As introduced at the end of the last chapter, this then introduces a host of other changes. The climate connections are many and complex, but in the end, as the Arctic loses its cryosphere, it also loses its soul.

EYES ON THE NORTH

At the NSIDC, where I have been the director since 2009, our most popular Web page is *Arctic Sea Ice News and Analysis*, which features daily tracking of the extent of the Arctic's floating sea-ice cover, and discussion about how current conditions compare to previous years.[1] It was developed in response to growing interest in the rapid decline in summer sea-ice extent over the period of satellite observations. Its audience ranges from fellow scientists to middle school students. In 2016 alone, *Arctic Sea Ice News and Analysis* had over 3 million visits. Questions from the public and media about what is happening to the sea-ice cover can be a near-daily occurrence and can be overwhelming in September as people focus on the daily-updated graph, anticipating the seasonal minimum in extent. Common questions include: "When will you 'make the call' on the minimum? Will it be a new record low? Why should we care about the loss of sea ice? Why wasn't there a new record this year?" NSIDC also gets it share of questions along the lines of "Why are you people perpetuating the myth of global warming?"

More recently, NSIDC started a companion Web site called *Greenland Ice Sheet Today,* which focuses on tracking daily changes in surface melt over the ice sheet, and discussing issues such as how summer melt extent is changing, and how it relates to the mass balance of the ice sheet and its contribution to sea-level rise.[2] As the extent of summer melt over the ice sheet increases, so does the popularity of the site.

NSIDC makes use of passive microwave data streams (more on this later) from the Defense Meteorological Satellite Program (DMSP) "F" series of satellites. In early 2016, NSIDC had planned to transition to the data stream from the F-19 satellite, the newest in the series, but F-19 suddenly died. Satellites do this sometimes. Then, as validation of Murphy's Law, one of the key channels on the older F-17 satellite that we had been using went wonky. So for several weeks, NSIDC had to suspend posting sea-ice maps and graphs and got an earful of questions and complaints. The Internet trolls claimed that we'd pulled the data because it wasn't showing what we wanted and were secretly cooking the books to show less sea ice. The trolls we can deal with. The big concern is that the DMSP F-18 satellite we've been using, already somewhat elderly, is the last in the series, and a suitable replacement might not be available for several years. A gap in coverage is a distinct possibility.

The NSIDC is by no means alone in trying to respond to the clamor for up-to-date information about the changing Arctic. The University of Illinois at Urbana-Champaign maintains its *Cryosphere Today* site.[3] The University of Bremen, Germany, provides updated maps of Arctic sea conditions based on the Advanced Microwave Sounding Radiometer-2 instrument carried on Japan's GCOM-W satellite.[4] The University of Washington provides updated estimates of Arctic ice volume through its Pan-Arctic Ice-Ocean Modeling and Assimilation System, better known as PIOMAS.[5]

Starting in 2006, the U.S. National Oceanic and Atmospheric Administration (NOAA) began issuing the Arctic Report Card, aimed at summarizing events of the year in the Arctic and updating trends.[6] It covers the gamut, with reports that include but are not limited to temperature and atmospheric circulation, sea-ice extent, ocean circulation, marine ecosystems, terrestrial ecosystems, snow cover, permafrost, glaciers and ice caps, the Greenland ice sheet, and river discharge. It is essentially a volunteer effort. Great care is taken to assure that only accurate information is conveyed. This is done through peer review organized by the Arctic Monitoring and Assessment Programme of the Arctic Council. The Arctic Council is an intergovernmental organization aimed at promoting cooperation, coordination, and interaction between Arctic nations, Arctic indigenous communities, and other Arctic inhabitants, especially on issues of sustainable development and environmental protection.

The Arctic Council has eight member countries: Canada, Denmark, Finland, Iceland, Norway, Russia, Sweden, and the United States. The United States was the chair from April 2015 through April 2017. The chair will be held by Finland until 2019.

Arctic nations (nations that border on the Arctic Ocean) have now fully woken up to the implications of the changing North, sometimes in interesting ways. In 2007, using two submersibles, Russia planted their national flag at the seafloor of the North Pole, with the intent of laying claim to the pole as its territory, hence laying claim to untapped reservoirs of oil and natural gas. Russia insists that the undersea Lomonosov Ridge, which basically extends to the pole, is part of the continental shelf of Siberia, and as such, Russia has sole rights to the ridge and the nearby seabed. While lauded by Russian President Vladimir Putin, the move was widely dismissed as puffery. In a more serious vein, on January 21, 2015, President Obama issued an executive order titled "Enhancing Coordination of National Efforts in the Arctic." To quote from Section 1, Policy: "The Arctic has critical long-term strategic, ecological, cultural, and economic value, and it is imperative that we continue to protect our national interests in the region, which include: national defense; sovereign rights and responsibilities; maritime safety; energy and economic benefits; environmental stewardship; promotion of science and research; and preservation of the rights, freedoms, and uses of the sea as reflected in international law."

While Obama didn't state it, the executive order is very much framed around the fact that the Arctic is losing its cryosphere. While the most relevant issue regarding the changing strategic and economic value of the Arctic is declining sea ice, making the Arctic more accessible to shipping, resource extraction, and other activities, this is, of course, only one aspect of the meltdown. So let's take a closer look at what is happening to the cryosphere, some of the drivers and impacts, and some other key changes. The best place to start is the Arctic Ocean and its sea ice, and how the loss of sea ice and other factors contribute to Arctic amplification. Then we'll turn to what is happening over land—permafrost and glacial ice—and then briefly visit fauna and flora.

SEA ICE

The most visible change in the Arctic cryosphere is the decline in sea-ice extent and volume. Sea-ice extent can be readily measured by satellite observations and is defined as the area of the ocean covered by ice with at least 15% ice concentration. Concentration refers to the fractional ice cover. For example, if a given satellite grid cell (pixel) is 30% ice and 70% open water, the ice concentration is 30%. Since late 1978, daily maps of ice extent have been compiled based on satellite passive microwave data—microwave radiation emitted from the surface. Microwave radiation is simply electromagnetic radiation in long (centimeter scale) wavelengths. By contrast, visible-band radiation that we can sense with our eyes (which is radiation emitted by the sun; we sense this radiation both directly and when it is reflected from surfaces) ranges from about 400 to 700 nanometers, a nanometer being 10^{-9} meters (one billionth of a meter). That's pretty short, so it's called shortwave (or solar) radiation. Different surfaces emit microwaves differently in different wavelengths and polarizations (vertical and horizontal), and the signal from sea ice is very distinctive. One of the great things about satellite passive microwave sensors is that, unlike systems that depend on visible-band radiation, the microwaves can be sensed day or night—remember that at the North Pole, it is dark for half the year. Also, microwaves of the type used to detect sea ice pretty much pass through clouds, which is important because the Arctic is generally a very cloudy place. A disadvantage of microwave sensors is the rather low spatial resolution. The resolution is sufficient to measure changes in sea-ice extent, but details can be missed.

The sea-ice extent waxes and wanes with the seasons. It is typically at its seasonal maximum in the middle of March; then the melt season sets in. The minimum extent occurs typically in the middle of September. The extent at different times during the year and the dates of the maximum and minimum ice extent have always varied from year to year. This is due to myriad things, including variations in air temperature, cloud cover, snow cover, and winds, as well as the transport of heat in the ocean that affects the growth and melt of the ice and how it moves (unlike some of my less-motivated undergraduate students, the ice doesn't just sit there like a stunned brook trout).

The sea-ice extent is trending downward for all months. Based on the period 1979–2016—essentially the full period of record from the multichannel passive microwave time series as of this writing, extent is dropping at about -3.2% per decade in January and about -3.5% per decade in February (with respect to monthly averages over the period 1981–2010).[7] The decline gets steeper from April through September. In September, which is the end of the melt season, the decline is a whopping -13.3% per decade. The September trend has gained the most attention, not just because it is the biggest, but because, as far as extent goes, it is the best reflection of the overall health of the ice cover. Basically, after all of the autumn and winter ice growth has occurred, and after all of the spring and summer melt has taken place, what's left? That would be the ice extent in September. The trend is smaller in the winter months, because even as things warm up (and they are), it is still plenty cold enough for ice to form in winter and extend to the shoreline, but that ice is fairly thin and prone to melting out the following summer.

The downward trends have by no means been even: superimposed upon the trends in each month are ups and downs that reflect the variations in weather and ocean conditions just mentioned. Some of these departures from year to year are quite large. The lowest monthly

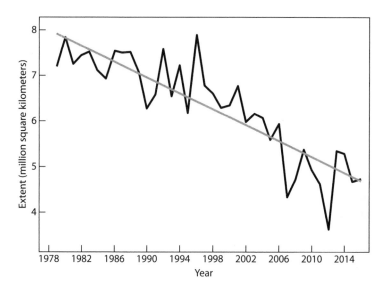

FIGURE 6. Average monthly September Arctic sea-ice extent from 1979 to 2016 and the linear trend line. Courtesy of National Snow and Ice Data Center (NSIDC).

averaged September extent in the record occurred in 2012, at 3.6 million square kilometers (fig. 6). Then there was a big jump the next year, and the September extent was down again in 2014 and 2015, with 2016 ending up very close to 2015. But the overall trend is compelling and can be further illustrated by comparing the mapped extent for three years: 1980 (near the beginning of the record, and the year that Ronald Reagan was elected to his first term as president), 1998 (partway through the record, and the year that President Clinton became embroiled in the Monica Lewinsky scandal), and 2012 (the record low year, when Obama began his second term as president; fig. 7). The September average extent for 1980,

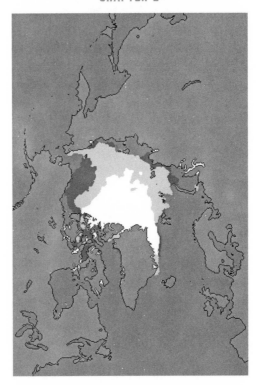

FIGURE 7. Extent of sea ice for September 2012 (white), 1998 (white and light gray), and 1980 (white, light gray, and dark gray). Courtesy of Walt Meier, National Snow and Ice Data Center.

of 7.8 million square kilometers, is approximately the area of the United States, minus Senator McCain's home state of Arizona. The extent for 2012, of 3.6 million square kilometers, is 46% of the area recorded in 1980. To make the same size comparison for 2012 would be like throwing out all of the states east of the Mississippi, all of the bordering states to the west, and then also tossing the Dakotas, Nebraska, and Kansas (fig. 8).

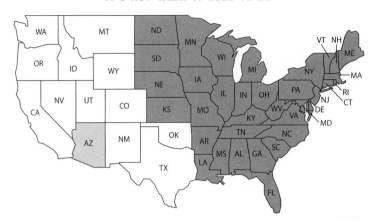

FIGURE 8. Sea-ice extent for September 1980 equaled the area of the contiguous United States minus Arizona. To make a comparison to September 2012, all of the dark-shaded states must be eliminated. Courtesy of Walt Meier, National Snow and Ice Data Center.

One would have to throw some of these states back in to make comparisons with the years 2013, 2014, and 2015, but the point is that the loss of September sea-ice extent over the satellite record is a lot of real estate.

There have been a number of efforts to extend the sea-ice time series to include years before 1979, even to as far back as 1850, using earlier forms of satellite data (back to 1972), and, before that, ship and aircraft reports. All of these analyses show the decline over the period of passive microwave coverage standing out prominently against the remainder of the record.

Ice extent does not tell the full story. To be complete, one needs to look at the volume of ice—that is, the product of the extent (in units of area) and the ice thickness (units of length, or depth). Volume, adjusted for ice density, in turn yields the ice mass, which has units of kilograms. Information for recent years is available from laser and radar altimeters aboard satellites and aircraft. Some longer records come from submarine sonar—submarines, traveling under the ice cover, operate upward-looking sonars. The first data came from the USS *Nautilus* in 1958, the first U.S. nuclear-powered submarine. The University of Washington Pan-Arctic Ice Ocean Modeling and Assimilation System (PIOMAS) mentioned earlier provides estimates of ice volume back to 1979. PIOMAS is based on a computer model that couples together the sea ice, the ocean, and the atmosphere and draws in many different types of observational data. While all of these data sources have their strengths and weaknesses, and computer models are never perfect, it is clear that ice volume (hence mass) is decreasing, not just from the decline in extent, but because the ice cover is also thinning. This thinning is driven by both a warming atmosphere and a warming ocean.

Because thickness varies so much from place to place and between seasons, the average thickness of ice is not the best measure of the thinning. Sea-ice scientists like to think more in terms of the statistical distribution of thickness. For example, upward-looking submarine sonar data collected back in April 1978 shows peak probabilities of ice thickness (that is, the most common thicknesses) between about 2.5 and 4.0 meters, but also thinner ice, right down to a veneer of a few centimeters thick or less. There is also a long tail in the statistical distribution of thick ice, up to 10 meters or even more. But over the years, the distribution has shifted toward thinner ice, and a lot of the really thick, resilient ice is gone.

Ice thickness is tied in with something called ice age. Ice that forms during a single growth season (e.g., it forms in October and then grows through the winter) is appropriately called first-year ice. Some of this first-year ice melts away during the summer, and some also exits the Arctic Ocean and enters the North Atlantic, primarily through the Fram Strait, the channel between northeastern Greenland and the Svalbard Archipelago. There it will eventually melt. The remainder—that which survives the summer melt season—becomes second-year ice. This ice will then tend to thicken through the next autumn and winter, mostly through growth at its bottom. Some of this second-year ice will also melt away the following summer (from both the top and the bottom) or be exported out of the Arctic Ocean, but again, some will survive to become third-year ice, which will again tend to thicken through the autumn and winter. Some of this may survive to become fourth-year ice and so on. The end result is that the Arctic Ocean has ice with a range of age classes, and in general, the older ice is the thicker ice. This is certainly not a hard rule, because first-year ice, when compressed by surrounding ice, easily crumples and forms ridges. The ridging process can make for very thick ice, and this, in turn, is the type of first-year ice that is more likely to survive that first critical melt season.

Ice age can be tracked using satellite data, and there is a record of ice age going back to 1985.[8] It used to be

A

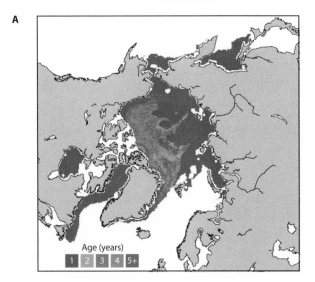

Age (years)

1 2 3 4 5+

B

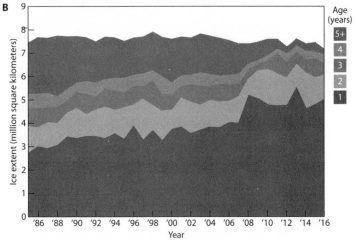

FIGURE 9. The distribution of sea-ice age classes for mid-March (week 11) of 2016 (top) and the change in age classes from 1985 through 2016 (bottom). Courtesy Walt Meier, National Snow and Ice Data Center.

that ice five years old or older was rather common, and some was over ten years of age, but with time, the ice has become progressively younger, with a pronounced loss of the oldest, thickest classes (at least five years old), which is consistent with other evidence for thinning (fig. 9).

ARCTIC AMPLIFICATION

An imperfect, but nevertheless convenient, metric of how much the Arctic has warmed in recent decades is the change in the air temperature near the surface. What emerges from such an analysis is that, for annual averages, the warming trend in the Arctic as a whole is about twice as large as the trend for the globe as a whole. Already introduced, this has a name—Arctic amplification. It is a complicated thing. First, Arctic amplification depends strongly on the season; overall it is strongest in autumn, winter, and spring, but it is largely nonexistent in the summer months. Second, the amount of warming varies greatly from place to place in the Arctic. Third, while sea-ice loss is a known driver of Arctic amplification, a number of other factors are involved.

Getting really good numbers on Arctic amplification is easier said than done, because the number of locations in the Arctic where temperature is directly measured at the surface is not nearly what is desired, but figure 10, which shows trends in the near-surface air temperature over the period 1979–2014 by season, based on an atmospheric reanalysis (see chapter 4), effectively summarizes what has been happening.

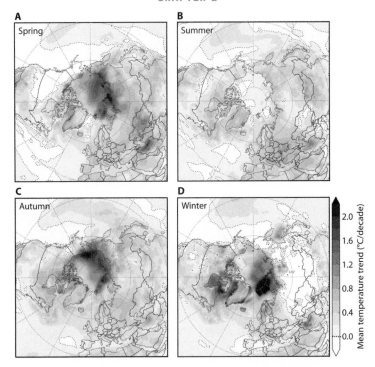

FIGURE 10. Trends in air temperature 2 meters above the surface by season over the period 1979 through 2014, expressed as degrees Centigrade per decade. Property of the author, created by Alexander Crawford at the National Snow and Ice Data Center.

During the autumn, the largest warming has been in the coastal seas north of Eurasia and Alaska that have also seen the biggest summer and early autumn declines in sea-ice extent. The Arctic is warming, and this is part of what is driving the sea-ice loss. As the climate warms, areas of dark open water form earlier in spring and summer than used to be the case, and these dark areas, with an albedo of about 10%, hence absorb 90% of the sun's shortwave radiation. However, when the sun sets in autumn, much of this energy (measured in units of joules) that was gained in about the top 20 meters of the ocean through spring and summer goes back to the atmosphere and keeps it warm (and then some of this is eventually radiated out to space). The atmospheric circulation (winds) will also tend to spread out the heat beyond the areas of sea-ice loss. Hence, while warmer conditions lead to loss of sea ice, the loss of sea ice itself boosts the autumn temperature change.

The patterns of warming for both winter and spring bear some resemblance to the autumn pattern. While by winter, sea ice has grown back along the coasts of Eurasia and Alaska, the ice is thinner than it used to be, and as such a fair bit of heat can be conducted from the ocean upward to the surface. The prominent blob of very strong winter warming (more than 2°C per decade) in the Barents Sea, on the Atlantic side of the Arctic, is consistent with the effect of relatively warm ocean waters entering the Arctic from the south, inhibiting winter ice formation. When winter ice covers a region, it effectively separates the fairly warm ocean from the colder atmosphere. Remove the sea ice and that ocean heat can now warm the atmosphere. This is part of what is happening in the Barents Sea in winter.

Albedo feedback also helps to explain part of the Arctic amplification in spring over land areas: there has been a trend toward less of a bright springtime snow cover, meaning that darker underlying surfaces are exposed, readily absorbing the sun's shortwave radiation and furthering the warming. By sharp contrast, Arctic amplification is fairly small over the Arctic Ocean in summer because, when ice is melting, the temperature right at the surface is fixed to the melting point of water, so the air immediately above the surface just can't budge much. Also, despite gaining a lot of energy, the temperature doesn't change radically over open water areas in summer because of the high heat capacity of water. To get a kilogram of water to increase 1°C in temperature, 4184 joules of energy need to be added. That's a lot of joules per kilogram, and even the top 20 meters of the Arctic Ocean contain a tremendous number of kilograms of water. But, as just stated, once the sun starts to set in the Arctic in autumn, this heat is released upward, seen as Arctic amplification.

As mentioned, there are a number of other factors linked to Arctic amplification, some of which remain to be completely understood in terms of their relative importance. Changes in cloud cover seem to be involved, as well as the observation that there is more water vapor in the Arctic atmosphere than there used to be (it has increased most sharply over areas of sea-ice loss).[9,10] The effect from having more water vapor in the atmosphere is related to the fact that water vapor, like carbon dioxide, is a greenhouse gas: it absorbs longwave radiation emitted by the surface, heats the other atmospheric gases by collisions, and emits longwave radiation back toward the surface. Clouds have a counterintuitive effect. Because of their high albedo, clouds limit the amount of solar energy received at the surface, but they also work like a blanket to reduce longwave radiation loss to outer space, and they are quite efficient at emitting longwave radiation back to the surface. For much of the planet, clouds have an overall cooling effect, but in the Arctic, except in the dead of summer, the blanket effect and longwave emission to the surface dominate, so they cause surface warming. There is evidence that the loss of sea ice has favored more cloud formation in autumn.

Another factor that leads to Arctic amplification is that the atmosphere near the surface does not readily mix with the air above, which traps the heat near the surface. This is because the Arctic is home to strong temperature inversions, in which temperatures increase upward from the surface rather than the more usual case of decreasing upward. At the top of the inversion layer in winter, typically 1000–1200 meters above the surface, temperatures might be 10°C or higher than those at the ground. This trapping is more or less the same thing that causes air pollution to get to dangerously high levels in the Los Angeles Basin, where inversions are also common. Changes in atmospheric circulation also play a role. For example, extreme heat waves over the Arctic Ocean observed in the winters of 2015/2016 and 2016/2017 were in part due to strong heat transport into the region from unusual patterns of atmospheric circulation. These storms also brought cloudy conditions and a lot of water vapor. Finally, there's a fascinating thing called the Planck effect, named after the famous physicist Max Planck. The longwave radiation emitted by the earth's surface depends on the temperature to the fourth power. As such, the increase in emitted radiation needed to balance a given radiative forcing (as imposed by higher greenhouse gas levels) requires a larger temperature increase at colder background temperatures. At 30°C, for example, an external radiative forcing of 1 watt per square meter (a watt is an energy flux of 1 joule per second) can be balanced by a 0.16°C warming, whereas at -30°C, a 0.31°C warming is required. The Arctic is colder than the tropics, so the Planck feedback also causes Arctic amplification.[11]

A final note regarding figure 10, as well as figure 6—why start in 1979? It turns out that 1979 is a very important year data-wise, especially for the Arctic. November 1978 marked the launch of the Global Weather Experiment, initially named FGGE, standing for the First GARP Global Experiment, GARP being the Global Atmospheric Research Program. FGGE is hence a nested acronym of the sort that scientists are unfortunately overly fond of adopting. The Global Weather Experiment was an audacious international venture to greatly improve earth observations. It represents the start of what is widely viewed as the modern satellite era, when the science community started to have a constellation of satellites in orbit to enable systematic monitoring of the environment, such as of sea ice. A number of important and related measurement programs were also started at this time, such as the Arctic Ocean Buoy Program at the

University of Washington, in Seattle, with the objective of deploying drifting buoys atop the Arctic Ocean sea-ice cover to measure air temperature, sea-level pressure, and ice drift. This later became the International Arctic Buoy Programme.[12]

PERMAFROST

While there is still plenty to be worked out regarding Arctic amplification, its influence on the Arctic environment cannot be denied. Consider the Arctic's permafrost. Temperatures measured in boreholes across the Arctic point to a general warming of permafrost over the past several decades, and in some areas near its southern limit, the permafrost is thawing. At least in Alaska, the permafrost warming rate increases to the north, partly because Arctic amplification also generally increases to the north. Another reason for the smaller warming in the south is that when thaw occurs, the change from ice to water takes up energy—and this is energy that would otherwise increase the temperature of the soils.

Permafrost thaw can have dramatic influences on the landscape, causing bucking and slumping of the ground, buildings, roads, and infrastructure, including pipelines. Much of the northern part of the Alaska pipeline that carries crude oil from the fields at Prudhoe Bay down to the southern coast at Valdez is built atop permafrost that is known to be warming.

FIGURE 11. Irina Overeem of the University of Colorado Boulder's Institute of Arctic and Alpine Research, a leading researcher on Arctic coastal erosion, standing on the eroding coastline near Drew Point, northern Alaska. Courtesy of Bob Anderson, Institute of Arctic and Alpine Research.

The interplay between sea-ice loss, warming, and permafrost thaw is central to the problem of coastal erosion. A good part of the Arctic coast is frozen sediments—permafrost. In decades past, when a summer storm came through, the presence of sea ice would limit how big ocean waves could get (the ice absorbs the wave energy), so coastal erosion was not a big issue. Now, with much less ice, winds have a long fetch over open water, resulting in larger waves and mechanical erosion. Because the ocean waters have warmed, the waves also result in thermal erosion because they hasten thaw of the permafrost. And the permafrost itself has also warmed. It's a triple whammy.

Irina Overeem and colleagues at University of Colorado's Institute of Arctic and Alpine Research have been looking at this problem in northern Alaska along the Beaufort Sea coast.[13] They estimate that the coastline midway between Point Barrow and Prudhoe Bay is eroding by 30–45 feet a year. The bluffs in this area, which are about 12 feet high, are comprised of frozen blocks of silt and peat. In summer, waves melt a notch at the bottom of the blocks that eventually undermines their bases. The frozen blocks of peat and silt then topple into the warm waves of the Beaufort Sea, which melts them in a matter of days and sweeps the remains out to sea (fig. 11).

GLACIAL ICE

As part of the 2013 NOAA Arctic Report Card, an international team headed by Martin Sharp of the University of Alberta, at Edmonton, conducted an updated assessment of mountain glaciers and ice caps outside of Greenland. Directly assessing mass balance using stakes and measurements of snow water equivalent (as was done at the St. Patrick Bay ice caps) is a very labor-intensive effort, but Sharp's team was able to analyze more than 20 glaciers with such direct measurements. Viewed collectively, there is an obvious pattern of mass loss since at least 1989 (fig. 12). While these results are from only a very small subset of Arctic glaciers, many

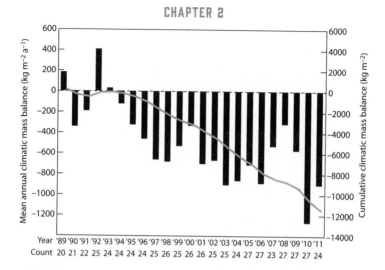

FIGURE 12. Mean annual (bars) and cumulative (heavy line) climatic mass balance from 1989 to 2011, based on all available annual measurements from Arctic glaciers reported to the World Glacier Monitoring Service by January 2013. Count refers to the number of glaciers analyzed. One kilogram per square meter represents a millimeter of water. Modified from Sharp, M., G. Wolken, M.-L. Geai, et. al (2013), "Mountain Glaciers and Ice Caps (outside Greenland)," Arctic Report Card: Updated for 2013, http://www.arctic.noaa.gov/report13/glaciers_ice_caps.html.

lines of evidence indicate that they reflect an Arctic-wide pattern, which in turn is part of a global pattern of mass loss from glaciers and ice caps. There is always an odd glacier or ice cap that is advancing or stable, but these are exceptions.

By the time this book goes to press, at least 30 different mass balance estimates of the Greenland ice sheet will have been conducted since 1998. Much of this work has relied on satellite and aircraft remote sensing; the sort of simple measurements conducted on the tiny St. Patrick Bay ice caps are completely unrealistic given the size of the ice sheet. Satellite remote-sensing tools include gravimetric measurements from NASA's Gravity Recovery and Climate Experiment (GRACE) mission; laser altimetry from NASA's Ice, Cloud, and Land Elevation Satellite (ICESat) and from its aircraft mission Operation IceBridge; and radar altimetry from the European Space Agency's Cryosphere Satellite (CryoSat-2) and Interferometric Synthetic Aperture Radar (InSAR).

GRACE is particularly fascinating. It makes use of two identical satellites, one following the other at a distance of about 220 kilometers. The local gravity field is mapped by measuring changes in the distance between the two satellites; the distances change because as each satellite passes over a gravitational anomaly, the gravitational pull is changed. Through repeat coverage, and with a lot of data processing, time changes in the gravitational field can be determined, primarily reflecting changes in terrestrial water storage. In the case of Greenland, these changes in water storage are related to ice loss. Altimeters and radars, by contrast, get at changes in the ice sheet elevation and ice thickness that are in turn related to mass balance. Ground-penetrating radars can get at the thickness of the individual big glaciers that drain the ice sheet. InSAR can also be used to obtain surface velocities of the ice sheet and the glaciers that drain it. As of late 2017, GRACE was on its last legs and running out of power, but a replacement is planned.

Satellite and aircraft information has been complemented by the application of regional climate models, as well as surface observations. Mass balance estimates range widely depending on the time period examined and the techniques used, but there is no doubt that the mass balance of Greenland has turned negative, meaning that the ice sheet is losing rather than gaining mass and hence is contributing to sea-level rise. This is due to both the calving of icebergs into the ocean (it was almost surely a Greenland iceberg that sank the unsinkable *Titanic* back in 1912) and surface runoff, the latter now appearing to have played the dominant role. The recent mass losses for Greenland are larger than for the Antarctic ice sheet, and the mass loss from Greenland seems to have recently accelerated. Back in 2012, glaciologist Andrew Shepard coordinated the input of many different scientists and past studies to come up with a best estimate of the mass loss.[14] The results are summarized in figure 13 as cumulative annual mass balance (gigatonnes) and equivalent sea-level rise (mm) for the Greenland and Antarctic ice sheets and for the two ice sheets combined. The accelerating mass loss from Greenland stands out. A gigatonne represents a cubic kilometer of water—that is, a cube of water with sides of 1 kilometer.

MORE WATER

The Arctic's water cycle is also faster now. Despite being rather small as oceans go, containing only about 1% of

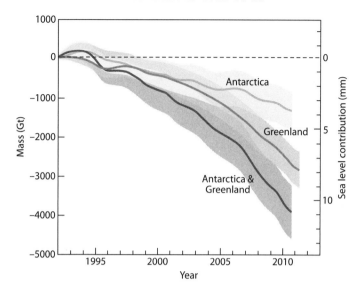

FIGURE 13. Cumulative annual mass balance (gigatonnes and equivalent sea-level rise) over the period 1992–2011 for Greenland, Antarctica, and Greenland and Antarctica combined. From Shepherd, A., E.R. Ivins, G.A. Valentina, et al. (2012), "A Reconciled Estimate of Ice-Sheet Mass Balance," *Science* 338. Reprinted by permission of AAAS.

the global ocean volume, the Arctic Ocean is special in that it receives about 10% of the global river discharge. The bulk of this discharge is contributed by a small number of large rivers, the four biggest being the Ob, Yenisey, and Lena in Russia and the Mackenzie in North America. Since the onset of reliable records in the mid-1930s, the discharge to the Arctic Ocean aggregated for the biggest Eurasian rivers has shown a general increase; changes over North America, for which records are only available from the middle of the 1970s onward, are also toward more discharge, although as with Eurasia, there is a lot of variability between different rivers and from year to year.[15] The changes can be readily explained only by an increase in net precipitation (and excess of precipitation over evaporation). A warming Arctic also implies greater evaporation, but the increase in precipitation seems to be winning.

The character of the river discharge is also changing. Due to melt of the snowpack that accumulates through autumn and winter, river discharge to the Arctic Ocean comes in a strong pulse in late spring and early summer. It then declines through summer and autumn to what is called the base flow. Because the climate is warming, the springtime pulse in discharge is creeping earlier into the season. This is consistent with satellite observations showing that, over the past several decades, the spring snow cover over northern lands has declined. According to snow scientist Chris Derksen of Environment Canada, over the period 1979–2015, the June snow cover extent over northern lands has decreased at the remarkable rate of -17% per decade. This analysis makes use of the same satellite passive microwave data stream used to monitor the coverage of sea ice.

FLORA AND FAUNA

Satellite data from the Advanced Very High Resolution Radiometer (AVHRR) sensor, available since 1982, has been used to assess photosynthetic activity using a measure known as NDVI, the Normalized Difference Vegetation

Index. The index is based on the way in which green plants reflect light in different wavelengths. Analysis of the NDVI data reveals that from 1982 through about 2012, there was a definite greening of the tundra, consistent with increased plant growth associated with a warming Arctic and a longer growing season. Since then, and for reasons that remain to be understood, the pattern has changed direction. Over the full period of record, there are also large areas showing a trend toward browning (again see the 2015 NOAA Arctic Report Card). There are indications that some areas have been undergoing a transition from tundra to shrub vegetation in response to a warming climate, a process known as shrubbification, but trends are not altogether clear, and there are large variations across the Arctic landscape.

The decline in sea-ice extent has been a big player in a substantial increase in primary productivity in the Arctic Ocean since the turn of the century.[16] Primary productivity refers to the rate at which carbon dioxide, either in the atmosphere or dissolved in the water, is converted by autotrophs (the phytoplankton that make their own food via photosynthesis) into organic matter. The phytoplankton get eaten by the zooplankton (little critters in the water), which are then eaten by other organisms, finally to the level of the top predators, a process commonly known as the food chain. The increase in primary productivity has been mostly along coastal areas near the shelf break (the edge of the continental shelves) where the loss of sea ice, which allows more sunlight to penetrate into the water, has also been attended by upwelled nutrients.

There have also been shifts in the population range of fish species. Marine biologists from the United States, Norway, Russia, and other Arctic nations have documented that species are moving poleward into the Arctic Ocean from both the Pacific and the Atlantic. This includes commercially important fish such as cod, haddock, and chinook salmon.[17] Research in the Bering Sea reveals that the species shifts can be linked to the combination of warmer ocean conditions, increased primary productivity in open-water areas that used to be ice-covered, the increasing abundance and biomass of Atlantic zooplankton, and lower fishing pressure due to fishing regulations.

In general, the public has shown little interest in zooplankton. Attention has instead tended to focus on the top of the food chain—polar bears, pinnipeds (walruses and seals), and whales of various species. The polar bear has become an icon of the changing Arctic and its shrinking sea-ice cover; many of us are familiar with the image of a polar bear standing on a shrinking ice floe. Polar bears spend most of their livelihood on the sea ice (but they mostly den on land); hence it follows that they will be adversely affected by loss of their habitat. Their primary prey is seals, which they hunt from the sea ice. It seems that polar bear numbers are overall in decline, attended by declines in body conditions, but the situation is complex. The International Union for Conservation of Nature Species Survival Commission Polar Bear Specialty Group, whose membership consists of scientists from nations signing the International Agreement on the Conservation of Polar Bears, concluded that of 19 polar bear populations examined, three (Baffin Bay, the Kane Basin, the Southern Beaufort Sea) are declining, six are stable, and one is actually increasing. For the remaining populations, data are insufficient to draw conclusions. However, there is widespread agreement that as sea-ice extent continues to decline, other bear populations will suffer. Bowhead whales, by contrast, seem to be doing pretty well. Bowheads annually migrate into the Beaufort Sea from the North Pacific via the Bering Strait. There they hang out in the Beaufort Sea to dine on zooplankton. Sea-ice loss has led to more favorable conditions for zooplankton, leading to improved body condition of whales.[18] There is subsistence hunting of bowhead whales by the Inupiat in Alaska, and this seems to be sustainable.

The situation with small whales that live in the Arctic—specifically, belugas and narwhals—is not well known. Narwhals (the males are readily identified by their long single tusk, which is an extended canine tooth) are known to be vulnerable to sea-ice loss, but it appears that they also have some flexibility when it comes to habitat selection. According to Donna Hauser of the University of Washington, beluga whales appear to be able to respond rather quickly to a changing environment, but whether different beluga populations will be able to adapt is unknown. Both species are widely hunted.

As outlined by Kit Kovacs and colleagues in their article in the 2015 NOAA Arctic Report Card, the situation with walruses and other pinnipeds is also complex, and it is difficult to separate effects of a changing Arctic climate from factors such as changing hunting regulations.[19] Walruses are widely distributed across the Arctic, and there are distinct North Atlantic and Pacific subspecies. There is also an isolated population in the Laptev Sea that many Russian biologists view as a separate subspecies. Walruses give birth on sea ice and use the ice as a haul-out platform for feeding, and for protection from storms and predators. In recent years, the public has taken notice of reports of particularly massive summertime walrus haul-outs on the land near Point Lay, Alaska, located on the shore of the Chukchi Sea. The first was observed in 2007, the year that at the time held the record for the lowest end-of-summer sea-ice extent. Walruses like to remain on the sea ice, which they use as a platform to dive to the ocean floor in search of mollusks and other prey. The summer sea ice now regularly retreats beyond the shallow continental shelf, so the walruses are obliged to haul out on land. An emerging problem is that the haul-outs have attracted tourists and the media, and all the attention can spook the walruses, causing a walrus stampede that can result in young animals being crushed and killed. Current thinking is that since walruses can haul out on land, loss of the sea ice by itself will not result in their extinction. However, fewer animals could likely be supported. Other concerns include ocean acidification that can affect the survival of their prey (calcifiers, like mollusks), the need for effective regulation of hunting, susceptibility to disease and contaminants, and effects of increased commercial shipping and oil and natural gas development.

Over the Russian Arctic coastal lands, there seems to have been an increase in rain-on-snow events during autumn and winter. This has been linked to sea-ice loss in the Barents and Kara Seas; all the open water makes the coastal areas warmer and also provides a moisture source for precipitation. Two major rain-on-snow events during winter, one in 2006 and another in 2013, are known to have led to a massive die-off of reindeer on the Yamal Peninsula (reindeer are close domesticated cousins of wild caribou). The basic link between these extreme events and the reindeer die-offs is that after it rains, the temperature drops, forming a hard ice crust that interferes with foraging. The 2013 event led to the death of 61,000 reindeer out of a population on the Yamal Peninsula of 275,000 animals.[20] One must always be careful in attributing individual extreme events to climate change; low-probability events happen sometimes. But their effect on the reindeer does not seem to be in dispute. For reasons that remain to be determined, the past decade has also seen a marked decline in caribou numbers over North America.

COMPLEXITY AND CONNECTIONS

As noted in chapter 1, the challenge scientists face in trying to understand the transformation of the Arctic is its astounding complexity. If carbon dioxide is added to the atmosphere, the Arctic warms, and sea ice and snow begin to melt. More solar energy is absorbed at the surface, and the extra energy gained in the upper ocean in spring and summer is released to the atmosphere in autumn and winter. But this is only part of the phenomenon of Arctic amplification. A warmer atmosphere holds more water vapor, but water vapor is also a greenhouse gas, furthering the warming, and the warming is also influenced by the strong vertical stability in the lower atmosphere and cloud cover. But changes in cloud cover can come about by changes in the amount of water vapor as well as changes in the circulation of the atmosphere, potentially driven by both changes in sea ice and events going on as far away as the tropical oceans! Rising temperatures and water vapor content can alter not just the amount and distribution of precipitation, but whether the precipitation falls as rain or snow. As it warms and patterns of precipitation change, the very vegetation of the Arctic lands changes, which by altering surface brightness and evaporation, also affects temperature and precipitation. Adding to the complexity is the inherent natural variability in climate on timescales ranging from weeks to months and sometimes even over spans of a decade or more, causing pronounced ups and down in things like temperature and sea-ice extent that are independent from changes in the greenhouse gas concentration. Even the most seasoned scientist can get dizzy trying to make sense of it all.

But when the Arctic began to stir in the early 1990s—by some accounts even earlier—it was not so much the complexity of the system that challenged our understanding; given how little data was available at the time, in many ways the challenge was simply documenting that things were actually changing. Only then could we begin the work of figuring out how the many different pieces of the puzzle fit together.

3

THE ARCTIC STIRS

The idea that an increase in the atmosphere's concentration of carbon dioxide would lead to warming and that, through feedbacks, the warming would be amplified in the Arctic, had been pretty well accepted by climate scientists well before I entered graduate school in 1982. Indeed, the basic issues had been outlined by the late 19th century. It was known that atmospheric carbon dioxide levels were on the rise, and experiments with even very early climate models were projecting a warming planet, with especially big changes in the Arctic. While the research papers I was reading as a student offered little observational evidence that anything had yet happened, it was natural that scientists started looking for signs of the anticipated warming. It was in 1990 that the Intergovernmental Panel on Climate Change (IPCC) guardedly concluded that, for the globe viewed as a whole, it might just be starting to emerge. Evidence that the Arctic itself was stirring then began to trickle in, but initially it wasn't convincing; records were short, and different studies often came to very different and even contradictory conclusions. With the passage of a few more years, the evidence firmed up. Still, it wasn't at all obvious why the Arctic was stirring; some of the emerging changes hinted at a human fingerprint, but much of it just looked like natural climate variability.

EARLY READINGS

To advance science, there must be written records that can be built upon. That is the role of publication in peer-reviewed scientific journals. Some of these journals go back to the 19th century and provide a fascinating avenue to assess the evolution of scientific thought. In the peer-review process, a scientist or a group of collaborating scientists writes a manuscript describing results from a research effort, and that manuscript is submitted to a journal in one of the professional societies aimed at that field of study. The paper is then sent out to be reviewed (also termed *refereed*) by other scientists with expertise in that field—that is, one's peers. Their job is to assess the paper's scientific merit and provide an assessment in a review, which may offer comments, criticisms, and suggestions to improve the manuscript, as well as a recommendation regarding suitability for publication. If, in the judgment of the reviewers and the journal editor, who has the job of coordinating the review process, the work meets expected standards (e.g., the techniques appear to be sound, the methods applied are appropriate, the conclusions that are drawn are supported by the data), the paper is published, almost invariably after revisions based on the reviews. Other scientists read the paper and may publish results from their own research (and present them at science conferences) that support or refute its conclusions. Peer review can be a slow process and is decidedly imperfect (it is, after all, a human process; egos, competition and envy can occasionally get in the way), but it works. In times past, there was the proud day when reprints from your paper came in the mail; you then walked down the hall, handing out autographed copies to colleagues. Nowadays everyone just downloads a copy. We've lost something in this regard.

At the library in February 1983, a few months before my second field season on the St. Patrick Bay ice caps, I was leafing through the bound periodicals for research papers on Arctic climate, and I came across a piece published in 1978 by Dr. John Walsh and Claudia Johnson.[1] John was then a postdoc at NCAR (the National Center for Atmospheric Research in Boulder, Colorado). The paper focused on variability in Arctic sea-ice extent for the period 1953–1977, using a largely pre-satellite sea-ice record painstakingly compiled from numerous sources, and examined how it compared to the existing record of Arctic surface air temperature. Through the 25-year record available to them, the authors found that the overall trend in ice extent was positive—that is, ice extent was on the increase, consistent with general cooling over that period. But the pattern was complex and puzzling. Above-average temperatures (with respect to the 25-year average) from 1959 to 1962 were followed by pronounced cooling and increased ice extent. Then, starting around 1965, the pattern reversed: temperatures increased, with a corresponding decrease in ice extent.

As John recalls it, "Climate had been cooling, especially in the Arctic. There were severe winters in the lower 48, and there was some talk, at least in the press, that we might be headed into an ice age. While at NCAR, I gridded monthly Arctic temperatures from land stations and Russian drifting stations and constructed a time series that showed Arctic temperatures starting to increase after about 1965. I showed this to Murray Mitchell and Will Kellogg, two famous climate scientists at NCAR, and their reaction was 'well, it looks like the Arctic has finally bottomed out and started warming like it should be.' Being a novice, I thought to myself, 'How can these guys be so confident that the Arctic should be warming?'"[2]

But as I had learned from Ray Bradley's classes, it had long been recognized that certain trace gases in the atmosphere, most notably water vapor, carbon dioxide, and methane, absorbed some of the longwave radiation emitted from the earth's surface, resulting in the natural greenhouse effect that makes our planet habitable. It followed that adding to the concentration of these trace gases, hence augmenting the greenhouse effect, would lead to further warming. Svante Arrhenius, who later won a Nobel Prize in chemistry, had conjectured on this way back in 1896. And measurements started in 1958 at the Mauna Loa Observatory in Hawaii showed that the concentration of carbon dioxide in the atmosphere was indeed increasing.

Missing was an understanding of how much warming there would be with a given change in the concentration of carbon dioxide along with other greenhouse gases such as methane. The science community, which I now fancied myself to be a part of, already knew a fair bit about feedbacks, such as the albedo feedback already discussed, and water vapor feedback (a warmer atmosphere holds more water vapor, but since water vapor is also a greenhouse gas, this leads to further warming), but we didn't sufficiently understand how different feedbacks interacted with each other, how the effects of atmospheric aerosols (very small particles or droplets that variously reflect or absorb radiation) entered into the mix, the influences of cloud cover, and the role of the oceans with their tremendous heat capacity.

But we made the best use of the tools that we had, and in 1979, Claire Parkinson, then a young scientist at the NASA Goddard Space Flight Center, teamed up with Will Kellogg to take an early look at how the sea-ice cover might respond to warming.[3] They made use of a sea-ice model that Claire had developed as a PhD student at NCAR with her adviser, Warren Washington, one of the true pioneers of climate modeling. "After completing the model," recalls Claire, "I started talking with Will Kellogg about the likelihood that an increase in greenhouse gases from human activities would lead to warming and whether this might affect the sea-ice cover. So we set up a fairly simple set of experiments in which we specified increases in temperature and changes in other boundary conditions, such as cloud cover. We found that in the face of a warming of 5°C, the Arctic ice cover in the model disappeared entirely in August and September but reformed in the winter."[4]

Just a year later, Suki Manabe and Ronald Stouffer of the Geophysical Fluid Dynamics Laboratory at Princeton University turned the problem around by using a climate model to estimate how much warming might actually occur[5]; this contrasted with Claire's experiment, in which an assumed amount of warming was simply supplied to the model. While the approach that Suki and Ronald took of quadrupling the concentration of carbon dioxide was extreme—they were basically hitting the climate system with a big hammer—they highlighted that the expected warming would be strongest over the Arctic Ocean and surrounding regions (6–9°C for annual averages, and bigger in winter), in large part due to the reduction in the thickness and extent of the floating sea-ice cover with its high albedo.

A remarkable effort came out the next year led by James (Jim) Hansen of the NASA Goddard Institute for Space Sciences.[6] This study is still relevant today and provides an insightful perspective on what was known about the climate system. Hansen addressed the global radiation balance, climate modeling, equilibrium climate sensitivity (how much the globally averaged air temperature at the surface will increase in response to a doubling of carbon dioxide), and how climate sensitivity depends on climate feedbacks. Jim also addressed issues of solar and volcanic forcing on climate (volcanoes can inject dust and aerosols high into the atmosphere) and the uncertainties related to aerosols and heat uptake by the oceans. Jim confidently predicted that the global average temperature would rise. He also came up with an estimated equilibrium climate sensitivity of 2.8°C for a doubling of carbon dioxide, a number that has withstood the test of time

Could anything be concluded from the study by John Walsh and Claudia Johnson of observed changes in Arctic temperature and sea ice over the period 1953–1977? Comments by John's colleagues at NCAR notwithstanding, the answer, in a word, is no. The data records were short, and the ups and downs in temperature and ice extent that they saw looked like natural climate variability. It was common knowledge that the earth's climate could vary quite a bit on many different timescales without any help from people. Saying that climate has natural variability is not to say that climate varies or changes without cause; everything has a cause. The natural cause might be external—for example, the sun starts shining a little brighter or dimmer. As already discussed, Milankovitch forcings can kick in climate feedbacks.

The short-term decadal-scale fluctuations, such as seen by John and Claudia, by contrast seemed like just the sort of thing that we now know can arise from purely internal processes in the climate system. For example, through feedbacks, small perturbations in the ocean surface temperature can reinforce themselves with lasting effects on the circulation of the overlying atmosphere and hence temperature.

So there I was, thinking about Arctic cooling and instantaneous glacierization, even secretly hoping for it, when the expectation for the future was just the opposite. Then again, observational evidence for warming had yet to be convincingly assembled, and climate science in general just wasn't something the public heard much about.

ENTERING THE FRAY

After finishing my project on the climate influences of the St. Patrick Bay ice caps in early 1985, I worked for a year and a half at what was then called the Lamont-Doherty Geological Observatory, located in Palisades, New York, in Rockland County, right on the New Jersey border. LDGO (now LDEO, as they view themselves as an environmental observatory) is part of Columbia University. I was a research technician, engaged, among other tasks, in mapping patterns of surface brightness over the Arctic sea-ice cover based on manual interpretation of data from satellite images. Rockland County is a great place to work and live if you have a lot of money, but I had very little. The attitude seemed to be that the privilege of working at LDGO was infinitely more important than being paid a living wage. Much of my diet consisted of Kraft macaroni and cheese.

I did learn how to program computers in Fortran. More importantly, I met my wife, Susan. Equally underpaid, she toiled downstairs as the assistant curator of the deep-sea core repository—a huge room filled with ocean cores from all over the world, used in paleoclimate research, including the ones used to confirm the Milankovitch theory of climate change. A bright point of our stay at Lamont was regular lunchtime campfires in the woods with the ever-engaging paleoclimate scientist George Kukla, featuring smoked sausage and overindulgence in cheap white wine.

Realizing that there really was no productive future for us at LDGO, I applied to the PhD program in geography at the University of Colorado Boulder and was accepted. That August, Susan and I hit the road to Boulder. We made a decision as to which of our cars would likely survive the trip and opted for my 1974 Impala, with its cracked windshield, mismatched doors gleaned from the junkyard, and retrofitted Chi-Chi Rodriguez hood ornament. We barely made it.

At Boulder, I ended up doing a study about how storms moving into the central Arctic Ocean in summer altered the circulation of the sea-ice cover, which then forced large openings in the ice. The project was funded by a grant from the Office of Naval Research (ONR). In the mid to late 1980s, ONR was funding quite a bit of work on Arctic sea ice and Arctic Ocean acoustics because it was relevant to submarine warfare. When ice grinds together, it makes a lot of noise under the water, and the big vertical variations in temperature and salinity in Arctic waters can confuse sonar. This combination helped the Soviets hide their submarines, and as such, the Navy wanted to know more about the Arctic Ocean. So my project, while unclassified and seemingly rather esoteric, had value to the ONR, as it informed our navy where and when the Soviets would be able to surface their submarines and shoot missiles at us. Or so I fancied. Regardless, it made for a good PhD dissertation, and I learned a lot about sea ice, climate, oceanography, and meteorology. I graduated in May 1989, eager to begin a career in Arctic science.

I turned down a postdoc position offered by the Byrd Polar Research Center at The Ohio State University in favor of staying on as a postdoc at the University of Colorado. I was given a clear message that one does not turn down a postdoc at the Byrd Center. However, Boulder, Colorado, seemed like a much nicer place to live climate-wise, and housing there had yet to become absurdly expensive. Furthermore, Boulder was a national hub of climate research, with the NCAR just west of town and NOAA labs just down the road from the university. Hopefully, the Byrd Center has forgiven my unintended slight.

SETTING THE STAGE

Just as I entered the PhD program at the University of Colorado Boulder in 1986, a paper came out in the journal *Science*, written by Arthur Lachenbruch and B. Vaughn Marshall, addressing changes in vertical profiles of permafrost temperature from boreholes in northernmost Alaska.[7] As they put it, "Because a temperature change at the surface takes time to propagate downward into the earth [via conduction], the deeper the temperature measurement, the farther back in time is the interval of the surface temperature history that it represents and the smoother is the signal as the high frequency [temperature] noise is progressively filtered out. Thus, the ground 'remembers' the major events in its surface temperature history. . . ." Phrased a bit differently, changes in vertical temperature profiles in permafrost provide information about the time history of surface air temperature, which gets around the relative lack of long-term measurements of air temperature in the Arctic. When Lachenbruch and Marshall analyzed the data, based on heat-conduction theory, they found that the change in the vertical profiles of temperature pointed toward a variable—albeit widespread—warming of the permafrost surface of 2–4°C during the last few decades to as long as a century. They stated, "Since models of greenhouse warming predict climate change will be greatest in the Arctic and might already be in progress, it is prudent to attempt to understand the rapidly changing thermal regime in this region."

Meanwhile, Phil Jones, Jim Hansen, and others had been busy trying to assemble the best possible time series of the global average surface air temperature based on station records. In 1987, Hansen and Sergej Lebedeff concluded that, for the globe as a whole, the period 1800–1985 had seen a global warming of about 0.5–0.7°C that was especially strong between 1965 and 1980.[8] Hence, the post-1940s global cooling observed especially in the Arctic, such as noted by Ray Bradley and others, seemed to be a fairly short-lived thing—a blip reflecting natural variability. There were certainly issues regarding how well the data-sparse Arctic was represented. Still, this was evidence that the climate was changing, lending support to conclusions of Arctic warming based on the changes in permafrost temperature in Alaska.

As various lines of evidence began to suggest that the planet was starting to warm up, momentum in climate research quickly grew. Climate modeling, already a hot topic, became hotter. While by today's standards, these models were simple and clunky, experiments from modeling communities around the world were consistently predicting that, as carbon dioxide levels rose, so would the global average temperature, and that the Arctic would be leading the way.

In 1990, the IPCC First Assessment Report was trotted out.[9] The IPCC had been established in 1988 by the World Meteorological Organization and the United Nations Environment Programme (organizations of the United Nations) to assess "the scientific, technical and socioeconomic information relevant for the understanding of the risk of human-induced climate change."

Some of the key statements (paraphrased a bit) from the report's executive summary are as follows:

Human activities are increasing atmospheric concentrations of greenhouse gases that will enhance the earth's natural greenhouse effect, resulting in warming of the surface. The main greenhouse gas, water vapor, will increase as a result, furthering the warming.

How much warming occurs through the 21st century depends on greenhouse gas emission rates, and warming will be attended by rising sea level.

Via climate feedbacks (especially albedo feedback), the polar regions will warm the most.

There are many uncertainties with regard to the timing, magnitude, and regional patterns of predicted climate change, due to an incomplete understanding of the sources and sinks of greenhouse gases and the role of clouds, oceans, and polar ice sheets.

The global mean surface air temperature has increased over the last 100 years, and while broadly consistent with predictions from climate models, it could still be largely due to natural variability. Unequivocal detection of the enhanced greenhouse effect is not likely for at least a decade.

So, while the IPCC was adamant that greenhouse warming would emerge, it was not at all certain whether we'd yet seen it. The section in the report on changes in the cryosphere gave no cause for alarm. There was evidence for a general recession of mountain glaciers around the world since the latter half of the 19th century, but the updated time series of Arctic sea-ice extent presented for the period 1972–1990 showed no trend. The report remarked that since about 1976, sea-ice extent had varied around a constant climatological level but that the 1972–1975 extent was significantly less. While some evidence was discussed on decreases in ice thickness between the years 1976 and 1987 based on submarine sonar data, it was acknowledged that the lack of a continuous data record made it impossible to know whether this was part of a long-term trend. A year before the IPCC report, Claire Parkinson and Don Cavalieri of NASA had examined the sea-ice record based on just the short satellite time series (1979–1987) from the Nimbus-7 Scanning Multichannel Microwave Radiometer (SMMR; it carried the first in the series of multichannel microwave sensors providing maps of ice extent on a daily basis) and found a very weak downward trend, and while I don't recall this study getting into the IPCC report, it would not have changed any of its conclusions. The short section in the IPCC report on permafrost noted the work of Lachenbruch and Marshall, but also argued that much of the inferred warming likely occurred prior to the 1930s and that since the 1930s there was little evidence for sustained Arctic warming.

But there were forward-looking minds. Responding to growing scientific interest in the Arctic and in climate change, in 1989 the NSF Office of Polar Programs (OPP) initiated the Arctic Climate System Study (ARCSS). The goals of ARCSS were to (1) understand the physical, chemical, biological, and social processes of the Arctic system that interact with the total earth system and thus contribute to or are influenced by global change, in order (2) to advance the scientific basis for predicting environmental change on a decade-to-centuries timescale and for formulating policy options in response to the anticipated impacts on humans and societal support systems.

So while it was called "global change" by NSF, a euphemism that persists in some circles even today, the direction that the NSF was taking was clear.

UNCONVINCING AND CONFUSING EVIDENCE

A couple of years after I graduated, in the spring of 1991, and having decided to try and make a career as a research scientist at the University of Colorado, an opportunity arose for more field work in the Canadian Arctic, this time on the sea ice in the channels near Resolute Bay. The effort was more or less led by a bunch of Canadian graduate students from the University of Waterloo, whose tolerance for cold was exceeded only by their tolerance for alcohol. I was there along with Jim Maslanik and Jeff Key, former office mates from the University of Colorado. We were researching variations in sea-ice thickness, springtime snowmelt, the development of melt ponds atop ice floes, and aerosol concentrations in the atmosphere, and were validating measurements of sea-ice surface temperature from satellite retrievals. We were also using a small (15 foot long) tethered blimp to see how the temperature varied at different heights above the surface.

Naïve enthusiasm got us into trouble when we inadvertently launched the blimp directly into the flight line of the final approach to the Resolute Bay airport. We had the blimp perhaps 1000 feet in the air, when, suddenly, the daily First Air Boeing 737 broke through the low cloud cover and screamed by, seemingly missing the blimp by a few feet. Tom Agnew, in charge of the blimp program, and the only seasoned scientist among us, looked at me and murmured, "oh crap, we are hosed." Though the small helium-filled blimp was probably never a threat to the Boeing twinjet, it was only moments later that the radio crackled alive with a booming voice from the airport demanding an explanation. It seems that the First Air pilot, upon seeing the blimp in his way, had become rather upset. We explained the situation, and Tom was brusquely summoned back to Resolute to discuss the matter. From what we were told, airport officials were yelling all the way down to Ottawa, which, for Canadians, is a very serious thing.

Less exciting but ultimately of greater scientific value were the series measurements taken with a sun photometer to assess the concentration of aerosols in the atmosphere. Basically, a sun photometer is a device pointed directly at the sun during cloudless periods (ours was of the handheld variety) that records the amount of solar radiation received at different wavelengths. Knowing the precise sun angle along with other information, one can determine how much the sun's beam is depleted by the aerosols. Remember that aerosols are small particles or droplets suspended in the air that, depending on their type, variously scatter or absorb solar radiation. We took a bunch of sun photometer measurements in 1991 with the intent of getting repeat measurements the next year.

On June 15, 1991, after we returned home, Mount Pinatubo in the Philippines blew its stack, injecting sulfate aerosols (the kind that scatter solar radiation) high into the stratosphere. When we returned to the field site in May 1992, we found a very different situation. It was much colder than in 1991, and there was no doubt that the eruption of Pinatubo was responsible. The eruption caused a decrease in the global average temperature from 1991 through 1993, and effects on the Arctic in the spring of 1992, at least where we were, were impressive. While sun photometer measurement clearly showed a big increase in aerosol concentrations relative to those taken in 1991, it was obvious even without them that sunlight was being blocked. You could see it from the color of the sky. The prospect of global warming seemed dim indeed.

Longtime colleague Chris Derksen of Environment Canada, then one of the students looking at the seasonal onset of snowmelt, still laments the cold spring of 1992. "I first entered the Arctic as a 20-year-old undergraduate student, under-informed, under-prepared, and under-equipped. Back then, it was still possible to walk into a professor's office to discuss a research assistantship, and a couple months later, despite a total lack of experience and expertise, be heading to the Arctic to live in a tent on the sea ice for three months. As exciting as the snow-mobiles, the snap and crackle of fracturing ice, and the odd polar bear encounter were, there was also a lot of tedium and repetition. Time moved slowly—days and weeks with the same companions, the same camp food, and very infrequent showers. My ticket home was the onset of the melt season. Only when melt was in full swing, and conditions on the ice became unsafe, could we decommission our research sites, break down camp, and claim victory. As impatient as I became, the melt season of 1992 had other ideas. With the eruption of Mount Pinatubo, the melt simply never came. Finally, as our food supplies, fuel, and patience dwindled, we had to pack up and leave, scientific tails tucked between our legs. No melt observations were to be made that season."[10]

My own skepticism regarding a warming Arctic was reinforced by a study that I published in January 1993 with Jon Kahl.[11] We looked at air temperatures over the Arctic Ocean measured at the surface and at several different levels in the atmosphere (up to the 700-millibar level; in the Arctic, this is typically somewhat less than 3000 meters above the surface), using air-temperature profiles from radiosondes and dropsondes. The radiosonde data came from the Russian North Pole program,[12] a series of drifting camps maintained on the ice cover, while the dropsonde data came from the U.S. Ptarmigan weather reconnaissance missions conducted by the U.S. Air Force. Collectively, these observations spanned the period 1950 through 1990. Radiosondes are balloon-borne instruments; they are launched at the surface and collect data on temperature, humidity, and winds as they rise. They are a key source of data ingested into numerical weather-prediction models used to make weather forecasts. Dropsondes are instead dropped from aircraft and collect data as they descend via parachute.

No evidence was found of the warming trends being predicted by the climate models of the time. The title of the paper, "Absence of Evidence for Greenhouse Warming over the Arctic Ocean in the Past 40 Years," is quite telling. To be clear, I was on board with the notion that the predicted warming would eventually emerge. What we said is that, based on the data, it hadn't happened yet.

Others shared similar views. As Jim Overland of NOAA, who has been doing Arctic climate research since the 1970s, put it, "In 1991–1992 we did not see or think much about Arctic changes. We did have the Manabe and Stouffer modeling paper showing that as carbon dioxide concentrations rose, we'd expect to see Arctic amplification. So we were on alert in that respect, but not a whole lot seemed to be happening."[13]

Marika Holland, a climate modeler and later senior scientist at NCAR with a longtime focus on Arctic sea processes, was just embarking on her career. "I began graduate school in 1992 and so was just starting to learn about the Arctic. I was reading papers by people like Suki Manabe illustrating that the Arctic should experience the largest warming in response to rising greenhouse gas concentrations. I was also starting work on sea-ice model development, and so previous studies like Manabe and Stouffer that discussed the power of the surface albedo feedback, were very intriguing. I honestly wasn't thinking too much about Arctic change, however. All of the previous work that I was studying discussed a future climate—not the one of today—and so it felt somehow far off and far away."[14]

Jen Francis, now a prominent researcher at Rutgers University, shared that feeling. "When I began graduate school at the University of Washington's Polar Science Center in 1988, there was little discussion of Arctic climate change. Even 'global warming' was still a fairly nebulous concept, and courses in climate science didn't exist [apparently Ray Bradley's course that I took was an exception]. It was well understood that increasing greenhouse gases would warm the earth, but observations of this phenomenon had not yet set off widespread alarm."[15]

And different lines of evidence for Arctic change seemed to be in conflict or served to confuse.

In January 1993, Bill Chapman and John Walsh took advantage of the longer sea-ice record then available, spanning the years 1953–1990, along with a gridded monthly record of Arctic air temperatures for the period 1961–1990.[16] The temperature time series made use of mainland and Arctic island stations, but provided no information from over the central Arctic Ocean. Like the study I was a part of that used radiosondes and dropsondes, they were testing what the climate models of the time were projecting—that as greenhouse gas concentrations rise, the increase in air temperatures will be greater in the Arctic than elsewhere, attended by a substantial retreat of sea ice.

Over the period 1961–1990, a warming trend was found, strongest over northern land areas during winter and spring, partly balanced by negative trends centered over the Atlantic-side subpolar seas. Over the period 1953–1990, Chapman and Walsh also found decreases in summer Arctic sea-ice extent, with new summer minima achieved three times in the last 15 years of the data record. No trend was found in winter ice extent. Their paper made a very bold statement: "The seasonal and geographic changes of sea-ice coverage are consistent with the more recent greenhouse experiments performed with coupled atmosphere-ocean models."

But wait a minute. We knew from John Walsh and Claudia Johnson's previous work that over the period 1953–1977 there had been a slight upward trend in ice extent, but that the more obvious feature was large variability, not just from year to year but from decade to decade. The IPCC report then made it clear that over the period 1972–1990 there had been no trend at all. But now, after looking at 1953–1990, there was a downward trend consistent with recent model projections? Given that it seemed to be the recent summers driving the trend, why didn't the IPCC analysis also show a trend? Perhaps the downward trend was just another example of natural variability in ice extent. There were also issues of how the data were processed and examined; while Chapman and Walsh looked at sea ice for different seasons, which yielded a significant decrease for summer months, the IPCC report and Walsh's earlier effort instead looked at standardized monthly anomalies. In computing standardized monthly anomalies, one first gets the long-term monthly averages (typically computed over the entire period of record or for some other selected base period), then subtracts the respective monthly mean from each of the individual monthly values in the record, and then adjusts by the monthly standard deviations. This approach removes the natural seasonal cycle in ice extent and puts the results for each month on a more even footing, but it can mask what may be important seasonally—dependent changes and trends.

What of the observed warming found over northern land areas? Why warming here but, from the earlier analysis using radiosondes and dropsondes, not over the ocean? Was it because the two studies were looking at different time periods? It had been clearly established from work such as Ray Bradley's that from around 1940 through at least part of the 1960s, there had been a cooling, so the warming noted by Chapman and Walsh was arguably inflated because their analysis began in a period known to have been fairly cold. Was there also anything about the quality of the data leading to different conclusions? Questions abounded. Adding to my own confusion, later in the same year, I was part of another effort looking at changes in tropospheric temperatures (temperatures above the surface) based on radiosonde data over land, spanning the period 1958–1986.[17] While considerable regional and seasonal variability was observed in temperature trends, no systematic changes were found. The punch line of the paper was very clear: "On the basis of our analysis, we conclude that greenhouse-induced warming is not detectable in the Arctic troposphere for the 1958–1986 period."

Of course, scientists continued to collect more data on Arctic conditions. With year-by-year lengthening of the satellite time series of sea-ice extent, especially when combined with earlier records, we slowly got a better handle on the natural variability in the sea-ice system, providing context to document any changes that might be emerging. Starting in 1979, the Arctic Ocean Buoy Program (which later became the International Arctic Buoy Programme) provided information on variability in surface air temperature, sea-level pressure, and ice drift over the Arctic Ocean with better spatial coverage than was possible from the North Pole program. We also started to get more data on Arctic oceanography. Mike Steele, an Arctic oceanographer at the University of Washington, in Seattle, and a longtime colleague, remembers the excitement of the times:

"I came to Seattle in 1987 to work with Dr. Jamie Morison as a postdoc after my PhD at the Geophysical Fluid Dynamics Laboratory in Princeton. At that point, Arctic oceanography was still extremely data-poor, and thus most papers were quite provincial. 'Ocean Conditions in Region X During Spring 1985' was a very typical sort of paper title. It reminded me of the 'What I did on my spring vacation' book reports from grammar school. There was rarely a big-picture context to these papers, and honestly, how could there be, with so few data points? But we did what we could do.

"And guess what? It turns out that the late 1980s and early 1990s was the start of a new era in Arctic oceanography, when research icebreakers started somewhat regular excursions into the deep basins to gather CTD [conductivity, temperature, and depth] and bottle data. For example, the *Polarstern* in 1987 went along 30°E longitude, reaching 86°N. In 1991, the *Oden* made it to the pole, the first one to do so, I think. Others soon followed, eventually on a nearly annual timescale. With these data, one could start thinking about the Arctic Ocean in a quantitative way. But at the same time, it was inevitable that people compared the new observations with the older observations."[18]

PROBING THE BRINY DEEP

It was with this growing database that the oceanographers began to suspect changes in what is known as the Arctic Ocean's Atlantic layer. The Atlantic layer is a relatively warm, salty layer that lies beneath a fairly fresh and cold surface layer of the Arctic Ocean. For most of the world's oceans, the warmest water lies at the surface, and the cooler water lies below. It is significant that the warmer water at the top is also the lighter (that is, less dense) water, which maintains a stable vertical profile, thus inhibiting vertical mixing. In the parlance of oceanographers, the *decrease* in temperature with depth, known as a thermocline, represents an *increase* in density with depth, known as a pycnocline. Think of homemade olive oil and vinegar dressing: even after vigorous shaking, it will separate so that the oil stays at the top, because the olive oil has a lower density. The vertical profile in most of the Arctic Ocean is particularly stable, but for a very different reason: at the low water temperatures found in the Arctic, the density is determined not by temperature, but largely by salinity. Simply put, the pycnocline is driven by a rapid increase in *salinity* with depth, known as a halocline. The water at the top is cold, which taken by itself should mean dense water, but this is overwhelmed by the fact that the surface water is relatively fresh, and in the Arctic Ocean, fresh means light. The fresh surface layer that is so important in forming this "cold halocline" is there in large part because every spring and summer there is a massive influx of freshwater into the fairly small and constrained Arctic Ocean from river runoff. Also, sea ice itself is fairly fresh (salt is rejected when the ice forms), so when it melts in summer, it freshens the surface waters. Also, on average, there is more precipitation over the Arctic Ocean than there is evaporation.

The underlying warmer but saltier water is brought into the Arctic Ocean from the Atlantic via two branches. The first and most important branch is through Fram Strait, which is the deep connection between the Arctic Ocean and the Atlantic, at about 78–80°N between the east coast of Greenland and the Svalbard Archipelago. This is called the West Spitsbergen Branch. As it enters the Arctic Ocean, just north of Fram Strait, this denser but warmer Atlantic water dives underneath the fresh, cold surface, forming the Atlantic layer. The other branch, through the Barents Sea, is appropriately known as the Barents Sea Branch. The Atlantic layer temperature maximum is a few degrees above the salinity-adjusted freezing point (salty water has a lower freezing point than freshwater) and typically lies between 100 and 800 meters deep. Below this is the deep water, cooler and somewhat fresher. Figure 14 shows a schematic of the Arctic Ocean's surface and subsurface ocean currents, and figure 15 shows typical profiles of salinity and density in the Arctic Ocean. The importance of the cold halocline is that, because it maintains such a stable density profile, it allows for sea ice to readily form in autumn and winter. Without it, the warm Atlantic waters would make it very hard for ice to form.

In 1991, German oceanographer Detlef Quadfasel and colleagues published a short letter reporting on measurements taken in August 1990 during a cruise of the Soviet nuclear-powered icebreaker *Rossiya*, which was carrying a group of wealthy tourists from Murmansk to the North Pole.[19] Detailed observations were made of ice conditions, and vertical temperature profiles were collected in the top 500 meters of the water column. Earlier data collected during 1987 showed that the maximum temperature of the Atlantic layer decreased from about 4°C in the ice-free Fram Strait region to below 2°C between Svalbard and Severnaya Zemlya, the archipelago

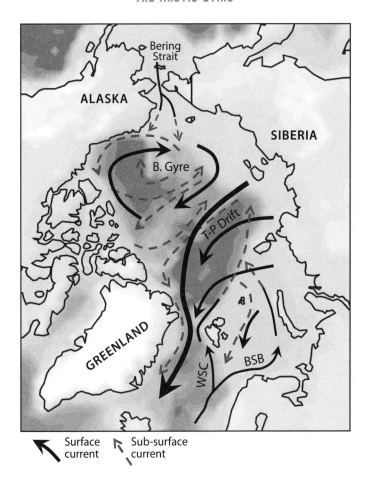

FIGURE 14. The circulation of the Arctic Ocean. Surface currents are shown by the solid arrows, subsurface currents by the dashed arrows. Gray shading indicates ocean depth (darker is deeper). Atlantic water enters the Arctic Ocean through the fairly deep Fram Strait between Greenland and Svalbard (the West Spitsbergen Current, or WSC) and through the shallow Barents Sea (the Barents Sea Branch, BSB) and then dives beneath the cold, fresh, surface layer. The major surface currents are the clockwise Beaufort Gyre (B. Gyre) and the Transpolar Drift Stream (T-P Drift). Property of the author, created by Alexander Crawford at the National Snow and Ice Data Center.

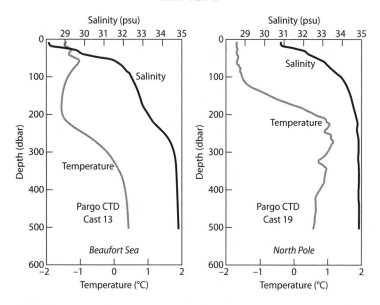

FIGURE 15. Relationship between salinity and temperature (x axis) and depth (y axis) for the Beaufort Sea and near the North Pole based on CTD (conductivity, temperature, depth) casts from the USS *Pargo* in the summer of 1993. The depth in decibars of pressure closely approximates depth in meters. Note that where salinity is increasing rapidly with depth (the halocline), the water temperatures are also cold. This means that the Arctic Ocean has a "cold halocline." Courtesy J. Morison, Polar Science Center, University of Washington at Seattle.

that lies to the east, along the path of the incoming Atlantic water now lying below the surface layer. The data collected in this region during the cruise of the *Rossiya* had a maximum temperature of 2.8°C, about a degree higher than values recorded in 1987. They also found that the ice thickness in the area was 20–30% less than expected.

As Detlef was quick to point out, a temperature difference between two years is statistically meaningless.

Sea-surface temperatures around Fram Strait were known to be variable, so the difference could just be random chance. The fairly low ice thickness might just be because the vessel, with its precious cargo of moneyed tourists, was taking the path of least resistance. They ended their letter stating that despite the sketchy nature of the observations, "It seems clear that there is a need for careful monitoring of the heat fluctuations in the Arctic, a region vulnerable to changes in the global climate."

As Mike Steele pointed out, more data was coming in from other expeditions, such as the cruise of the *Oden* in 1991. The *Oden* was a fairly new Swedish icebreaker, built in 1988. During 1991, the *Oden* and the German research icebreaker *Polarstern* became the first non-nuclear surface ships to reach the North Pole. The *Oden* data, analyzed by an international team of oceanographers, showed a slight but definite warming of Atlantic-derived waters near the pole compared to earlier data, suggesting that the warming identified by Quadfasel was real and might be fairly widespread.[20]

Then, in 1993, came the Arctic cruises of the USS *Pargo* and the *Henry Larsen*, followed in 1994 by the *Polar Sea* and the *Louis S. St. Laurent*, in turn followed in 1995 by the USS *Cavalla*. Data from all of these cruises basically pointed to an increased influence of Atlantic-derived waters and warming of that inflow. The USS *Pargo* and *Louis S. St. Laurent* data also showed puzzling warm cores of water over the Lomonosov and Mendeleyev Ridges with Atlantic-layer temperatures greater than

1.5°C (warm for the Arctic!) for which there was no evidence in earlier records. Apparently, Quadfasel was onto something.

The USS *Pargo* and USS *Cavalla* are of particular note as they are United States Sturgeon Class nuclear attack submarines (*Pargo* and *Cavalla* are the second submarines with those names; the first were *Gato*-class vessels serving in the Second World War).

The 1993 *Pargo* cruise was the first of what was to become a series of six Scientific Ice Expeditions (SCICEX) conducted between 1995 and 1999, collecting various types of hydrographic data and also providing detailed information on sea-ice thickness from upward-looking sonar. Since 1999, a modified approach has been taken, with time set aside for the collection of unclassified scientific data during otherwise classified submarine exercises. For the *Pargo* cruise (fig. 16), a scientific staff of seven joined the Navy crew; the science staff included Jamie Morison of the University of Washington, in Seattle, who was to soon become a prominent voice raising attention for the need to understand the oceanographic changes seemingly unfolding in the Arctic.

SHIFTING WINDS

To summarize where things stood around the year 1993, while evidence was growing that sea-ice extent was

FIGURE 16. USS *Pargo* (SSN-650), September 1993. Source: US Navy Arctic Submarine Laboratory.

on the decline in summer, the picture was very muddy: the changes were not very big, and a lot depended on the time period looked at and the techniques used to analyze the data. Temperature trends seemed to depend not just on the season and region analyzed and the time period of analysis, but also on whether one was looking at temperatures at the surface or aloft. Despite growing suspicion that something was going on in the Arctic Ocean, nobody knew whether this was a temporary change—part of some natural cycle—or the harbinger of something bigger.

Right around this time, shifts in atmospheric wind patterns were noted. In 1993, I led a study with Jason Box, Roger Barry, and John Walsh examining patterns of cyclone and anticyclone activity in the Arctic, using a record of sea-level pressure extending from 1952 through 1989.[21] Cyclones are the low-pressure systems that residents of middle and higher latitudes are familiar with, associated with warm fronts, cold fronts, and precipitation. In the Northern Hemisphere, surface winds blow anticlockwise around cyclones. Anticyclones are just the opposite; they are the high-pressure systems typically associated with sunny weather, with clockwise winds at the surface in the Northern Hemisphere. The study of cyclones and anticyclones is germane to climate science, because the winds associated with these weather disturbances are one of the primary mechanisms through which heat and water vapor are transported from the warm lower latitudes into the colder higher latitudes. We found that for the Arctic as a whole, cyclone numbers had increased over the 1952–1989 period during winter, spring, and summer, and anticyclone numbers had increased during spring, summer, and autumn.

The meaning of these changes was entirely lost on us. A big concern was that the cyclone and anticyclone statistics recorded for the earlier years of the record might not be very reliable due to scanty observations. Hence, the trends might be just a result of changing data quality rather than a real climate signal. However, we noted that the changes, if real, might bear on John Walsh's earlier finding that surface air temperatures had increased over land areas, primarily in winter and spring; more cyclones and anticyclones suggest a stronger poleward transport of heat and water vapor, which would be consistent with increases in air temperature. On the other hand, the work with radiosondes and drop-sondes pointed to the absence of temperature trends.

Recall that in 1979, the Arctic Ocean Buoy Program started providing information on the circulation of the sea-ice cover from drifting buoys, as well as measurements of temperature and sea-level pressure. The pressure measurements were in turn leading to more accurate maps of sea-level pressure over the Arctic Ocean and hence the pattern of surface winds (the distribution of pressure drives the speed and direction of the winds). It was not until February 1996 that a comprehensive quantitative analysis using this data was conducted by John Walsh, by now one of the major players in Arctic climate science.[22]

John showed that over the period of record then available, spanning 1979 through 1994, the annual average sea-level pressure over the central Arctic Ocean had decreased. Furthermore, for every calendar month, the annual mean pressure was lower in the second half of the 1979–1994 period than the first, but with the strongest changes in autumn and winter. This decrease in sea-level pressure was manifested in weakening of the mean high-pressure area (the anticyclone) centered over the central Arctic Ocean, meaning that the circulation of the winds had become more counterclockwise, or cyclonic. These changes were compensated for by increases in pressure at lower latitudes over the subpolar oceans.

At least from a broad-brush point of view, this seemed consistent with what we had seen in terms of cyclone numbers; if the number of cyclones was increasing, this would imply a reduction in the average pressure over the Arctic for the simple reason that cyclones are themselves low-pressure systems. But we had also seen increases in anticyclones, so it was not clear how the two studies connected. One sure thing was that the quality of the data that John was looking at was better than what we had been working with.

Rather telling of the uncertainty of the times, the shifts in pressure documented by Walsh were very much framed around the idea of natural decadal-scale variability (such as documented in a number of other studies over different parts of the world) rather than some sort of indirect response to greenhouse warming, such as sea-ice loss altering the surface energy budget, which then influences the circulation of the atmosphere.

As John recalls, "We intentionally downplayed the anthropogenic influence in that paper. At the time, we were aware of several modeling studies that had shown decreases of Arctic sea-level pressure in response to pre-scribed sea-ice loss [i.e., an ice-free Arctic Ocean], and we mentioned those modeling studies in the 1996 article. However, as of the mid-1990s, there had been little loss of sea ice, so it was hard to link the sea-level pressure change to sea-ice changes, whether anthropogenically driven or otherwise. Hence we chose to emphasize internal [nat-ural] variability in the interpretation."[23]

John was unable to pin down a cause of the pressure changes. However, the study was important as it led to recognition that there was probably a link between the changes in the surface winds accompanying the changes in sea-level pressure and the changes being observed in the ocean. Winds strongly drive the surface circulation of the ocean, and through a process known as Ekman pumping, these circulation changes might reach deeper ocean depths with time. Also, if the amount of exported sea ice (such as through Fram Strait) or Arctic near-surface waters varies over time due to wind shifts, there could conceivably be a connection to deep-ocean convection in the Greenland–Labrador Seas region having subsurface impacts.

Just a few months later, in June 1996, I finished a little study with my former graduate school office mate Jim Maslanik.[24] We were looking at changes in sea-ice extent over the period November 1978 through September 1995. Though we made a point of saying, "No clear Arctic-wide response of the sea-ice cover to global climate change has yet been identified," we also argued that recent extreme September sea-ice minima in the satellite record, such as seen in 1990, 1993, and 1995, could be linked to a sharp increase since 1989 in the frequency of low-pressure systems (cyclones) over the central Arctic Ocean.

Our conclusions regarding sea-ice/cyclone links would have changed had we waited a few months and incorporated data for September 1996. Though passing with surprisingly little notice, September sea-ice extent for 1996 ended up being the highest in the satellite record. To this day, it remains the highest in the record, and is almost certain to remain so. It was also a very stormy summer.

So why did September 1996 end up with a lot of sea ice, when we had argued that the low September extent seems to follow summers with lots of cyclones? In hindsight, it was probably a case of just not having enough data, or not looking at it in the right way, to really pin down what was going on. Subsequent work, such as by Masayo Ogi, then at the University of Washington, made it clear that stormy summers generally (but not always) favor more sea ice at summer's end. A stormy summer pattern tends to be a cold pattern, which limits the summer melt. The wind pattern also favors a diverging ice cover because the generally counterclockwise (cyclonic) winds act to spread the ice out to cover a larger area. This sounds counterintuitive, because winds converge at the surface under a low-pressure system, but if you add up all of the forces on the ice, at least in summer conditions, a cyclonic wind pattern favors ice divergence. Yes, our paper passed a rigorous peer review, but just because a paper passes peer review does not mean that the conclusions are correct. And we were incorrect.[25] So while including 1996 would have blown our postulated sea-ice/cyclone link out of the water, it would only have strengthened Walsh's finding of a recent trend toward lower sea-level pressures over the Arctic Ocean.

What neither John Walsh, Jim Maslanik, nor I were immediately aware of, probably because it was happening at about the same time, was the work of Jim Hurrell, then a young scientist at NCAR (as of this book's writing he's the director) focusing on the North Atlantic Oscillation (NAO).

THE RISE OF THE NAO

It has been said that even the Vikings knew about the NAO. That is likely false. As discussed by Thomas Haine of Johns Hopkins University, in an article in *Weather* magazine (a publication of the Royal Meteorological Society), the Vikings knew a great deal about the North Atlantic environment but had neither the data nor the means to understand what the NAO was all about. Nevertheless, influences of the NAO have certainly been recognized for centuries. The first written account comes from the diary of the 18th-century Danish missionary to Greenland Hans Egede Saabye, published in 1745, in which he states that the Danes were well aware that when winters in western Greenland were especially severe, the winters in Denmark tended to be mild, and vice versa.[26]

The NAO reflects a co-variability between the strengths of two semipermanent "centers of action" in the atmospheric circulation in the North Atlantic—the Icelandic Low, so named because it is centered near Iceland, and the Azores High, a high-pressure cell centered roughly over the Azores. This co-variability means that when the Icelandic Low is strong (its central pressure is especially low, meaning lots of cyclones in the area), the Azores High is especially strong (its central pressure is especially high). This is called the positive phase of the NAO. In the negative phase, both the Icelandic Low and the Azores High are weak. What this also means is that when the NAO is in its positive mode, there is a big gradient in pressure between the Icelandic Low and the Azores High. A big pressure gradient means strong winds, with a component from west to east, which in winter brings lots of storms and warm and moist air into northern Europe, leading to mild and wet winters. However, on the back side of the Icelandic Low, winds have a component from the north, leading to cool conditions in the area west of Greenland (fig. 17). When the NAO is in its negative mode, the pressure gradient between the Icelandic Low and the Azores High is weaker, and the storm track shifts south. Northern Europe is cooler, but the area west of Greenland is fairly warm.

The first formal analysis of the NAO was back in 1932, by Sir Gilbert Walker. It quickly became apparent that one could form a simple index of the phase and strength of the NAO from the normalized (that is, departures from a long-term average that are adjusted for standard deviation) difference in sea-level pressure between a station in Iceland and a station in the Azores.

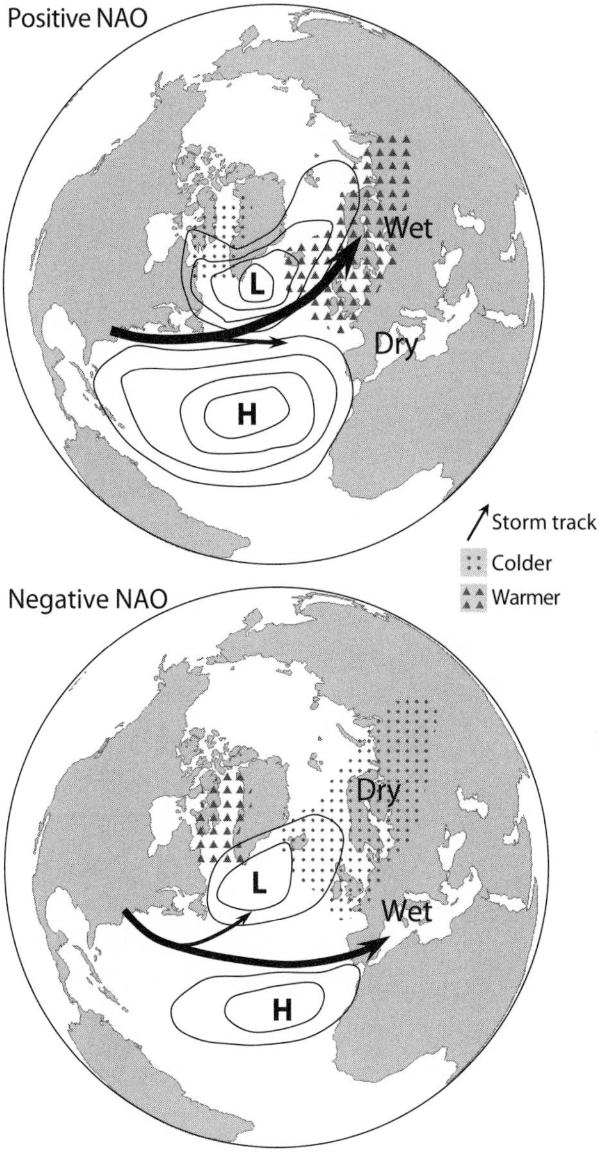

FIGURE 17. Positive (top) and negative (bottom) phases of the North Atlantic Oscillation. Property of the author, created by Alexander Crawford at the National Snow and Ice Data Center.

There are a number of variants of this station-based index, but they all get at the same thing: when the normalized pressure is negative (below average, or low) over Iceland and positive (above average, or high) over the Azores, the NAO is in a positive mode, and the magnitude of the pressure difference is a measure of the strength of the positive NAO pattern. The opposite holds for the negative NAO phase.

In August 1995, Jim Hurrell reported that over the previous decade, the winter NAO had been largely stuck in its positive mode, broadly consistent with a pattern of recent warmth across Europe and cold conditions in the northwestern North Atlantic.[27] In a subsequent study that came out in March 1996, barely a month after Walsh published his work, Hurrell expanded on this topic, finding that nearly all of the winter warming since the 1970s at the surface across Europe and even downstream over Eurasia (hence over a bigger area than depicted in fig. 17) and nearly all of a compensating cooling over the northwestern Atlantic over the same time could be accounted for by a trend in the NAO.[28] Specifically, the winter index of the NAO had risen from primarily negative index values in the 1960s and 1970s to positive index values peaking in the early to middle 1990s, meaning a progressively stronger Icelandic Low and a stronger Azores High.

This was of extreme importance because it provided immediate insight into the patterns of temperature change that Chapman and Walsh had noted earlier. It also showed that influences of the NAO were more widespread than previously appreciated. It certainly didn't explain everything. For example, Hurrell's analysis couldn't say what was going on temperature-wise over the data-sparse Arctic Ocean. But his studies were seminal in providing insight into why different studies of temperature change based on different time periods were coming up with different conclusions regarding trends—what you got depended, in part, on what the NAO had been doing over the period of record being examined.

I recall being sobered by these results. With so much depending on the NAO, how could one hope to identify any signal of greenhouse warming? Jim Hurrell pointed out what while some elements of the observed temperature anomaly patterns since the 1970s (not only those in the higher latitudes) resembled the greenhouse-warming fingerprint that was being predicted by climate models, it was difficult to say whether the observed changes were a response to greenhouse warming or just part of a natural decadal-scale variation in the atmospheric circulation. Might the change in the NAO also help to explain the changes in sea-level pressure over the Arctic Ocean noted earlier in Walsh's February 1996 paper?

Jim Hurrell recalls his thinking at the time. "For good reason, a lot of discussion and attention was being paid to the similarity of the observed changes in surface temperature and those projected by climate models in response to enhanced greenhouse gas forcing. Several articles were emerging that were claiming that the changes—or at least elements of them—were beyond the range of natural variability. Yet, my analysis of changes in sea-level pressure and related changes in atmospheric circulation were highlighting decade-long shifts in major patterns or modes of climate variability. Kevin Trenberth of NCAR and others were already looking at changes over the Pacific, but the Atlantic changes, related to the NAO, had not received much attention. When I was able to quantify the spatial pattern and magnitude of the changes in surface temperature due to the NAO, and the contribution of those changes to the overall hemispheric warming, it became clear to me and others that more research was needed to understand the mechanisms of internal variability on decadal and longer timescales, and how modes of variability like the NAO might be affected by anthropogenic change. It is interesting to me that these same questions are still being examined today!"[29]

DEAR COLLEAGUE AND THE IPCC

Only five years had passed since Detlef Quadfasel's initial oceanographic observations, and only three years had passed Chapman and Walsh's paper on sea ice and temperature, but by 1996, there was broad consensus that the Arctic was indeed changing. We were far from understanding why. Was this some type of natural variability in the Arctic system that may well have occurred before but wasn't detected because we didn't have the data? The NAO link found by Hurrell and the circulation changes noted by Walsh and others could be interpreted as supporting this view. Were we seeing some of the emerging signals of greenhouse warming? The broad agreement between some of the changes and what was being projected by climate models supported this view. Maybe it was a combination of the two.

In my view, it was around 1996 that the science community really started to organize to try and figure things out, both in small groups and in larger groups, via e-mail and in person. It was largely the oceanographers who initially led the way. In December of that year, a group of eight prominent oceanographers, including Jamie Morison, Mike Steele, James Swift, Knut Aagaard, Miles McPhee, Kelly Falkner, Robin Muench, and Norbert Untersteiner, along with climate scientist John Walsh, drafted a "Dear Colleague" letter on the issue of Arctic change; it was widely distributed via e-mail to the Arctic research community with the intent of galvanizing support to study the problem. It started with this statement: "This open letter is the first step in development of a program to track and understand major changes in the Arctic environment. The program is tentatively called The Study of Arctic Change."

Given that the letter was largely written by the oceanographers, it naturally focused on changes in the Arctic Ocean. Links with shifts in atmospheric circulation were also discussed, drawing on the work of Walsh and colleagues and additional research by Jamie Morison that was just about to be published. Notably absent was a discussion of confusion regarding sea-ice trends and the conflicting evidence for changes in air temperature. In this sense, the interpretation that there had been "major changes in the Arctic environment" was perhaps still a matter of debate.

The letter also emphasized that observations might represent decadal-scale change (to my mind, this was the best interpretation of the evidence) but could also represent the start of a longer-term shift associated with greenhouse warming. Why Hurrell's work on the NAO wasn't mentioned in the "Dear Colleague" letter is unclear. Most likely it was simply overlooked, given the broad oceanographic focus and the many different studies that were being published at the time. The letter went on to argue the need for an oceanographic-measurement program with surveys, drifting buoys, and ocean moorings, and that this would have to be an international program. It also recognized the need to better understand atmospheric variability and assess the evidence for past events like the shift toward low pressure documented by Walsh.

The Second IPCC Assessment Report had been published earlier that year.[30] Along with an assessment of observed changes, it contained a great deal of discussion about growth rates of greenhouse gases, radiative forcing on climate, climate modeling, and projections of climate change. The Summary for Policymakers stated,

"The balance of evidence suggests a discernable human influence on global climate." This was based not just on analysis of the observations, but on advances in detection and attribution aimed at distinguishing between human-induced and natural influences on climate. These included (1) comparisons between the observed climate time series and proxy (paleoclimate) records to help put the present into perspective, (2) the use of climate models to better understand the magnitude of natural variability versus "forced" climate responses (climate responses to increased greenhouse gas concentrations), and (3) pattern-based studies, whereby observed patterns of change are compared to those simulated from climate models forced by observed changes in greenhouse gases and aerosols.

The IPCC report cautioned that the ability to quantify a human influence on climate was limited because the expected signs of this influence were still emerging from the noise of natural climate variability. In this sense, the report agreed with Hurrell's arguments with respect to the NAO as well as the tone of the "Dear Colleague" letter highlighting the possibility this was just decadal-scale variability. Pointing to ongoing struggles regarding interpretation of the data, the IPCC stated that neither hemisphere had exhibited significant trends in sea-ice extent since 1973. It spoke to the notion that sea-ice trends were sensitive to the season and time period examined and to how one processed the data. The IPCC also concluded that sea-ice thickness, as measured by submarines and based on upward-looking sonar, was variable, but at least for the period 1979–1990, had shown no trends. The report did, however, acknowledge evidence that Northern Hemisphere snow cover had decreased in recent years (1988–1994), especially in spring.

So, looking back, 1996 was a watershed year. The IPCC Second Assessment Report had been published, with stronger albeit still guarded statements about a role of humans in climate change. The "Dear Colleague" letter came out and the Arctic research community was coming together.

But in many ways, events over the next five years only added to uncertainty as to where the Arctic was headed.

4

UNAAMI

By the middle of 1997, barely six months after the "Dear Colleague" letter had been written, it seemed that just about every scientist who knew where the North Pole was had become enamored of the changing North. The oceanographers prepared for a major field program called the Surface Heat Balance of the Arctic Ocean (SHEBA). Climate modelers took charge of trying to predict how the Arctic was going to evolve over the coming century. Scientists like me who loved getting their hands dirty with data took advantage of progressively lengthening climate records along with new sources of information. Peoples of the North were becoming partners in the quest to understand. What would the future bring? Will the Arctic of old return? If not, then how will we cope?

A new way of looking at the atmosphere quickly overshadowed recognition of the North Atlantic Oscillation as a major driver of change. In the eyes of many, the North Atlantic Oscillation began to be seen as just the little sister of something bigger and more fundamental called the Arctic Oscillation, or AO. The AO, it seemed, provided a more complete explanation of what was happening across the Arctic, and scientists from far and wide jumped onto the AO bandwagon. The Study of Environmental Arctic Change (SEARCH) was born, and the AO framework became a centerpiece. But what were we to call this increasingly coherent set of Arctic changes? Another stilted acronym wasn't going to cut it. The term *unaami* was offered, a Yup'ik word for "tomorrow," reflecting the uncertain future.

But at the close of the 20th century, the science was still far from settled, and confusion reigned. By many accounts, an Arctic response to the ever-growing greenhouse gas levels in the atmosphere should have emerged. However, the problem of attribution remained. Many of the environmental records were still too short to come to firm conclusions. Some were of uncertain quality, and others didn't provide enough spatial coverage to be able to say what was going on for the Arctic as a whole. From an analysis of the temperature record, it also seemed that the high-latitude warming, albeit impressive, was actually no bigger than the interdecadal temperature range during the 20th century. Unaami, strongly linked to the AO, still largely looked a lot like natural variability.

BUILDING MOMENTUM

The SHEBA field effort, sponsored by the National Science Foundation and the Office of Naval Research, got underway on October 2, 1997, when the Canadian Coast Guard icebreaker *Des Groseilliers* came to a halt in the Beaufort Sea. The icebreaker was allowed to be frozen in, beginning a yearlong drift that lasted until October 11, 1998. At any given time, there were between 20 and 50 researchers at Ice Station SHEBA. Don Perovich, of the Cold Regions Research and Engineering Laboratory in Hanover, New Hampshire, was the lead scientist. He recently recalled, "SHEBA was the first experiment I was involved with that focused on climate change. However, SHEBA was not about detecting change. It was focused on understanding the ice albedo feedback and cloud radiation feedback and using that understanding to improve the treatment of Arctic sea ice in climate models."[1] While SHEBA was a milestone because of the data collected (data that is still being used today), arguably its greater contribution was in further galvanizing the Arctic research community. Whether or not they participated in the field program, hundreds of scientists played a role in SHEBA. The program brought people together, which is always the key in moving science forward.

It was also in 1997 that improved information on patterns of atmospheric circulation started to become available from "atmospheric reanalysis" efforts. A reanalysis basically involves assimilating all of the historical atmospheric information one has (such as from radiosondes, satellites, dropsondes, aircraft reports, and sea-level pressures from drifting buoys) into a numerical weather model. It is the same basic approach used in weather prediction systems that form the basis of TV weather forecasts, but the difference is that a reanalysis uses a fixed version of the weather model. In operational weather forecasting, modelers are constantly tweaking the computer code to get better forecasts. But in an atmospheric reanalysis, tweaking is not a good idea. The goal of a reanalysis is to get internally consistent atmospheric fields spanning multiple decades. Any small change made to the weather model changes the output, and this can result in confusing jumps through time in things like temperature and sea-level pressure that are unrelated to climate; they are just a result of the tweaks. A reanalysis is designed to avoid this problem. Reanalysis data suddenly made it much easier to examine long-term variability in atmospheric circulation patterns and see how recent changes (such as those documented by John Walsh and Jim Hurrell) fit into the bigger picture. The first reanalysis was a joint effort between the National Centers for Environmental Prediction (NCEP, part of NOAA) and NCAR, known as the NCEP/NCAR Reanalysis.[2] The European Centre for Medium-Range Weather Forecasts (ECMWF) soon followed with their own effort.

However, a longer-term perspective was still needed: how big were these recent Arctic changes in the context of the past several hundred years? Eyes turned to the paleoclimate community, and that same year, a team led by Jonathan Overpeck (then at the University of Colorado Boulder) used various proxy records to assemble a 400-year record of Arctic summer temperatures. Their major conclusion was that the 20th-century Arctic was the warmest of the past 400 years.[3]

Jonathan and his colleagues tried to explain the variability in their record in terms of three components: the contributions from changes in greenhouse gas concentrations, changes in the amount of radiation emitted by the sun (irradiance), and aerosol loading from volcanic eruptions. Despite the many uncertainties in the analysis, they concluded that a period of Arctic warming spanning the period 1820–1920 was primarily the result of reduced volcanic activity and increasing solar radiation. After 1920, high solar radiation and low volcanic aerosol loading likely continued to have an influence, but growing greenhouse gas concentrations probably played an increasingly dominant role. Overpeck's study didn't prove that greenhouse warming had reared its head in the Arctic, but it certainly got more people leaning in that direction.

Turning back to observational records, it was recognized that we didn't have much direct information on surface temperatures over the Arctic Ocean. Was it warming over the ocean or not? Walsh and others' studies couldn't say because they were limited to analyzing data collected over mainland areas and Arctic islands.

My work with Jon Kahl using radiosondes and drop-sondes focused on the atmosphere above the surface. In 1997, Seelye Martin at the University of Washington, in Seattle, piloted an effort to see what was going on over the Arctic Ocean using the Russian North Pole records for the period 1961–1990.[4] The reason they didn't include more recent years was simply that the North Pole program ended in 1990 with the breakup of the Soviet Union. Though these records provided only sparse spatial coverage (just a few drifting stations were in operation at any given time), Seelye found some surface warming over that period, mostly in May and June. This contradicted what had been found using the dropsonde and radiosonde data, which showed no trends. Though Seelye made some arguments to account for the discrepancy, the issue was left unresolved. Data from the new NCEP/NCAR reanalysis also didn't show very much going on over the Arctic Ocean; what stood out instead was the pattern of warming over land areas and cooling over the northwestern North Atlantic associated with the NAO trend.

To further understand shifting atmospheric patterns, in 1997 I was part of another effort looking at cyclone activity over the Northern Hemisphere.[5] We were able to show that the recent reductions in sea-level pressure over the Arctic Ocean, noted a year earlier by Walsh, could be associated with increased cyclone activity over the area. We also extended an earlier analysis demonstrating that over the analyzed period, 1966–1993, there had been an increase in cold-season cyclone counts over the Arctic as a whole and a decline in counts in lower latitudes. This seemed to link with the change in the NAO discussed by Jim Hurrell and also hinted at something bigger, such as an overall poleward shift in storm tracks.

The period around 1997 and 1998 also saw a great deal of modeling work, especially on decade-scale climate variability. Andrey Proshutinsky and Mark Johnson (both at the University of Alaska Fairbanks at the time) showed that the Arctic Ocean had two opposing "regimes" of circulation forced by changes in wind patterns—a cyclonic (counterclockwise) regime and an anticyclonic (clock-wise) regime, each persisting for 5–7 years.[6] They argued that these contrasting regimes could help to explain the changes in Arctic Ocean hydrography that had been observed, as well as variability in sea-ice conditions. The behavior of the Beaufort Gyre was central to the argument. Recall from figure 14 that the clockwise Beaufort Gyre is one of the big features of the sea-ice and upper-ocean circulation in the Arctic. The clockwise (anticyclonic) Beaufort Gyre is in turn driven by the average clockwise circulation of the overlying atmosphere, known as the Beaufort Sea High. During the anticyclonic regimes, the Beaufort Gyre is strong. During the cyclonic regimes, the Beaufort Gyre is weak, and ice motion near the pole has a more counterclockwise component. Andre and Mark speculated that changes in sea-surface temperature in the far North Atlantic initiated transitions between the two regimes, which then triggered shifts in atmospheric circulation. This thinking was in line with ideas at the time that there might be "internal oscillators" within the Arctic giving rise to decadal-scale climate variability, explaining changes in atmospheric circulation such as seen by John Walsh and colleagues. Lawrence Mysak of McGill University was a big proponent of this basic idea. These shifts and the explanations offered for them seemed to have nothing at all to do with anthropogenic influences on climate.

Along similar lines, in May 1998, Mike Steele and Tim Boyd of the University of Washington showed that the cold halocline in the Arctic Ocean's Eurasian Basin had retreated during the 1990s to cover significantly less area than in previous years.[7] Steele and Boyd emphasized that because the cold halocline acts as an insulator, preventing the Atlantic-layer heat from mixing upward, its retreat could have a big effect on the sea-ice cover and, more broadly, the Arctic's energy budget. They surmised that the observed retreat of the cold halocline might be related to a shift in surface wind patterns that influence the location where freshwaters from the rivers flow into the deeper parts of the Arctic Ocean.

As Mike relates, "I was plotting up various ocean layers [Atlantic Water, various halocline layers, surface waters] and comparing them to past literature from Aagaard, Carmack, Jones, and Anderson. I recall making such plots and laying them all out on a conference room table, just staring at them. And lo and behold, I find that the recently observed spatial patterns and amplitudes of these are very different from what I expected. First thing is to double-check that I'm doing it right. Yes, looks okay. Then the inescapable conclusion is that these layers have shifted relative to past observations. Wow, so at the same time that I'm just trying to get a good pan-Arctic map, I've got to deal with change as well. Nice. At the end of my cold halocline retreat paper, I cited Proshutinsky and Johnson with regard to climate oscillations. But I also noted that some recent observations had values outside of the range of historical observations. So perhaps something more than a simple oscillation was happening."[8] So, while greenhouse warming need not be invoked at all to explain the cold halocline retreat, Mike seemed to be leaving the door ajar.

THE RISE OF THE ARCTIC OSCILLATION

On May 1, 1998, just a few weeks before Mike Steele's work on the retreat of the cold halocline came out, Dave Thompson, then a graduate student at the University of Washington, in Seattle, published a paper with his adviser, Mike Wallace, titled "The Arctic Oscillation Signature in the Wintertime Geopotential Height and Temperature Fields."[9] It had an immediate and lasting impact on the direction of Arctic research. In many ways, the research directions were highly productive. In others, as we will see, the ensuing Arctic Oscillation mania and bandwagon was perhaps a bit of a red herring.

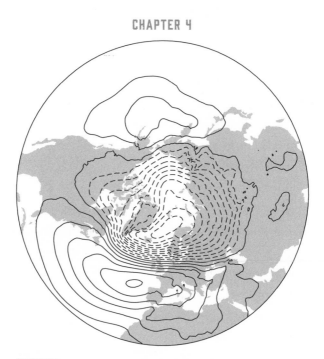

FIGURE 18: The Arctic Oscillation pattern at the surface. The Arctic Oscillation has three centers of action; the North Atlantic Oscillation has only two. Source: Hurrell, J. W., Kushnir, Y., Ottersen, G. and Visbeck, M. (2003) "An Overview of the North Atlantic Oscillation," in *The North Atlantic Oscillation: Climatic Significance and Environmental Impact* (eds. J. W. Hurrell, Y. Kushnir, G. Ottersen and M. Visbeck), American Geophysical Union, Washington, D. C.

Dave and Mike basically argued that the NAO should be viewed as a regional expression—just a piece—of a bigger mode of atmospheric variability, which they termed the Arctic Oscillation, or AO, which also came to be known as the Northern Annular Mode, or NAM. Much like Pluto, which after many years of respect was demoted by astronomers to the embarrassing status of a dwarf planet, the NAO was demoted from a fundamental atmospheric mode to something less. The winter AO pattern is depicted in figure 18. This is based on something called an Empirical Orthogonal Function (EOF) analysis of surface pressure fields, which we don't have to get into to understand the basic issues (I never fully understood EOFs myself, which is perhaps why I've never particularly cared for them).

The contour lines on figure 18 show three bull's eyes, which correspond to the AO "centers of action" in the atmospheric circulation at the surface. There is an Arctic center (the dashed-line contours) and two lower-latitude centers (the solid contours), one of them over the North Atlantic and a weaker one (fewer contours) over the North Pacific. This is contrary to the NAO, which has only two centers of action corresponding to the Icelandic Low and the Azores High. In the AO way of thinking, if surface atmospheric pressures are higher than average over the Arctic center, they are lower than average over both the Atlantic and Pacific centers (lower over both because the Atlantic and Pacific centers have the same sign, indicated in figure 18 by both of them being depicted with solid contour lines). Conversely, if surface pressures are lower than average over the Arctic center, then they are higher than average over both the Pacific and Atlantic centers. Like the NAO, the phase and strength of the AO can be described in terms of an index value, but this case is based on the fancy EOF analysis. The change through time in the winter index of the AO was found to exhibit the same general trend that Jim Hurrell had documented for the NAO—a switch from generally negative values in the 1970s to positive values in the 1990s. This meant a general shift from the combination of high pressure over the Arctic Ocean and low pressure over the Atlantic and Pacific centers (High Arctic Ocean, Low Atlantic, Low Pacific) to its reverse (Low Arctic Ocean, High Atlantic, High Pacific).

Thompson and Wallace argued that changes in the phase of the winter AO are controlled by what is going on not at the surface, but higher up in the atmosphere—namely, in the stratosphere. The stratosphere is the layer of the atmosphere that starts about 5 miles above the surface in the Arctic and about 6–8 miles above the surface in the middle latitudes. The stratosphere is the second major layer of the earth's atmosphere; the troposphere lies below it, and the mesosphere lies above it. Temperatures within the stratosphere increase the higher one goes. This is because ozone in the stratosphere absorbs ultraviolet radiation, and the ozone concentration also increases upward in the stratosphere, peaking at around 15 miles in altitude.

The fundamental feature of the circulation in the stratosphere is something called the circumpolar stratospheric vortex—a large-scale counterclockwise (cyclonic) motion of the winds. Within the vortex, the winds blow around the planet pretty much along lines of latitude from west to east. The circulation is hence termed zonally symmetric, or annular (ring-like). By contrast, lower down in the troposphere, the circulation pattern is wavier—the winds blow primarily from west to east but with north and south meanders, a bit like one sees in a big river. Closer to the surface, the circulation becomes even wavier and typically breaks up into individual whorls that we see as cyclones, anticyclones, and their associated warm and cold fronts.

The basic argument advanced by Thompson and Wallace is that if the circumpolar stratospheric vortex becomes stronger for some reason—that is, the stratospheric winds blow stronger from west to east (as atmospheric scientists would say, the vortex "spins up")—the AO pattern at the surface, such as seen in figure 18, moves toward its positive phase, with surface pressures lowering over the Arctic center of action (seen as a shift toward stormier, more cyclonic weather) and increasing over the Atlantic and Pacific centers of action (seen as a shift toward nicer, more anticyclonic weather). If the stratospheric vortex weakens for some reason (it spins down), the AO at the surface moves toward its negative phase; surface pressures rise over the Arctic and fall over the Atlantic and Pacific.

Hence, what happens in the stratospheric circumpolar vortex and what happens to the AO pattern at the surface are intimately connected to each other, or in the parlance of atmospheric scientists, the stratosphere and the surface are coupled. And it is the change in the stratospheric vortex that controls the switch in the phase of the surface AO from positive to negative or vice versa. A great deal of work following Thompson and Wallace's paper focused on how the switch in the AO phase could involve processes such as "sudden stratospheric warmings" that can rapidly change the stratospheric vortex.

The rise of the AO paradigm turned conventional thinking on its head. Even before Jim Hurrell had documented the trend in the NAO, scientists had been pondering why the NAO would sometimes get "stuck" for extended periods in its positive or negative phase. Many scientists thought it was the effects of forcing by slowly varying Atlantic or tropical sea-surface temperature anomalies. They argued that big persistent blobs of unusually high or low sea-surface temperatures could influence heating of the overlying atmosphere, which could then influence the atmospheric circulation far away from the heating influence, manifested in a change in the phase of the NAO.

By sharp contrast, in the AO paradigm of stratospheric control, scientists focused on losses of stratospheric ozone or increasing levels of tropospheric carbon dioxide, because both act to cool the stratosphere. Cooling then spins up the stratospheric vortex, seen at the surface as a shift toward the positive phase of the AO. Loss of ozone means less absorbtion of ultraviolet radiation. Oddly enough, while an increasing concentration of carbon dioxide leads to warming at the surface and throughout a deep layer of the atmosphere, it causes the upper stratosphere to cool.

Another important argument for the winter AO being "fundamental" is that it has a counterpart in the Southern Hemisphere, called the Antarctic Oscillation or Southern Annular Mode. However, the Antarctic Oscillation has a simpler appearance—it does not have three centers of action but is rather largely symmetric, or annular—if pressures are lower than average over the Antarctic, they are higher than average to the south, and vice versa. As the argument goes, the AO at the surface would look a lot more like its better-behaved Southern Hemisphere cousin if not for factors in the Northern Hemisphere that distort the atmospheric flow, such as the Rocky Mountains, the Himalayas, and the high Greenland ice sheet, and because in winter, it is warmer over ocean than land. The Antarctic Oscillation pattern is more symmetric for the simple reason that there are fewer distorting influences in the Southern Hemisphere; apart from South America, winds have a pretty clear shot around the Antarctic continent. So in the Southern Hemisphere, one can view a shift toward the positive phase on the Antarctic Oscillation as basically moving atmospheric mass from Antarctic latitudes (surface pressure drops) to more northern latitudes (surface pressure increases). The same idea holds in the Northern Hemisphere—a shift toward the positive AO phase shifts atmospheric mass away from the Arctic into the mid latitudes, but the shift in mass to the mid latitudes is focused in the Pacific and Atlantic centers shown in figure 18.

It was found that, like the NAO, the change in the winter AO, expressed as the broad positive trend in its index value, could explain the recent pattern of surface warming across Europe and downstream over Eurasia, and the compensating cooling over the northwestern Atlantic. However, compared to the NAO, the AO seemed to give a more complete explanation of temperature patterns (fig. 19).

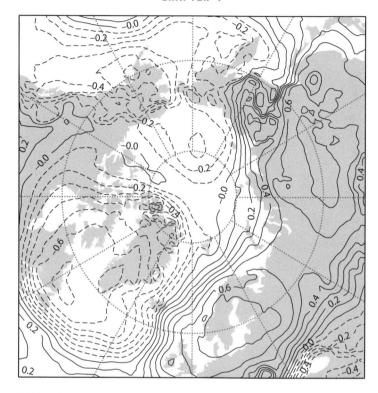

FIGURE 19. Spatial correlations between the winter Arctic Oscillation index and the surface air temperature. Bigger numbers mean a stronger relationship. When the AO is in its positive phase, temperatures are above average (as shown by the solid contour lines) over much of Eurasia and below average (as shown by the dashed contour lines) for the region centered over northeastern North America and Greenland, as well as over the North Pacific. The AO is not strongly correlated with temperatures over the central Arctic Ocean. The temperature data are from an atmospheric reanalysis. Property of the author, created by Alexander Crawford at the National Snow and Ice Data Center.

The pressure changes over the Arctic noted by Walsh also seemed to fit better into the AO framework than into the NAO framework. Furthermore, to many, the physics behind the AO were appealing. While there was by no means universal acceptance of the AO framework in the science community (more on this later), it quickly became a dominant paradigm, and it did so at a critical time.

THE DAWN OF SEARCH

Roughly six months earlier, in November 1997, a large open workshop, supported by NSF ARCSS, was held at the University of Washington, in Seattle, with the intent of exploring the changes taking place and designing a science program to study and understand them. It had a strong proponent in Mike Ledbetter, who headed the ARCSS program, and it reflected momentum that grew from the "Dear Colleague" letter that had been disseminated the previous year. Presumably, some of the scientists had to leave the SHEBA Ice Camp to attend the workshop.

The workshop was titled The Study of Arctic Change. There were presentations spanning the ocean, atmosphere, and ice, based on both observations and modeling studies. Working groups were assembled, and there were breakout sessions to address research questions and develop research strategies. Excitement was palpable. The Arctic was changing, there was a big field program going on with SHEBA, and the community had the ear of the NSF as well as other agencies, like NASA and NOAA. It was a good time to be an

Arctic scientist. The workshop report was published in August 1998.

By the time of the workshop, the AO had become the talk of the town, and through presentations and discussions it became better understood that the changes unfolding in the Arctic Ocean and the atmosphere were broadly related to each other. Viewed most simply, and I italicize for emphasis, *changes in surface wind patterns related to the shift toward the positive phase of the winter AO were driving changing patterns of surface air temperature, as well as sea-ice and ocean circulation, expressed in many of the hydrographic changes observed in the Arctic Ocean, potentially including the stronger inflow of Atlantic waters and retreat of the cold halocline.* But there were many questions. How clean was the AO link? Was there also a background warming that could explain the emerging downward trend in sea-ice extent? The paleoclimate work from Jonathan Overpeck and colleagues, indicating that summer temperatures over the Arctic were the highest in the past 400 years, suggested that there might be. Was the change in the winter AO index just an expression of natural decadal-scale variability, or might it reflect some sort of anthropogenic influence, such as stratospheric cooling?

Scientists love mysteries. NSF was sold on The Study of Arctic Change, and a few months later, on October 20, 1998, the new program was given the slightly different name Study of Environmental Arctic Change, or SEARCH, basically because adding the extra word (environmental) made for a catchier acronym. It should be obvious by now that scientists are also notorious for coming up with acronyms for everything, sometimes double- and even triple-nested acronyms. Sometimes the acronym comes into such wide use that the original words that made up the acronym become irrelevant (e.g., NASA) or are even forgotten. But SEARCH wasn't bad as far as acronyms go.

With some funding from the NSF, a SEARCH science team was assembled to write a SEARCH Science Plan. I was asked to be a member of the SEARCH Science Steering Committee. We met in April 1999 to write a preliminary outline (all big NSF science projects need a science plan) and put together an invitation list for a Science Plan workshop. The original membership in attendance, in addition to myself, included some scientists already introduced: Jamie Morison (the SEARCH Chair), Jim Overland (he and Jamie were the big movers and shakers), and Jonathan Overpeck, along with other prominent voices, including David Battisti (atmospheric dynamics, University of Washington), Hajo Eicken (sea ice, University of Alaska), and the refreshingly irreverent Lou Codispoti (Arctic Ocean biogeochemistry, University of Maryland; Lou had little patience for matters of bureaucracy). As SEARCH gained traction, additional members were added to address biology (Jackie Grebmeier, University of Tennessee) and the social sciences (Jack Kruse, University of Massachusetts).

To help frame the Science Plan, Jamie Morison wanted "a word to capture what I felt was a coupled ocean atmosphere behavior of decadal scale. The idea was to keep things focused and give a name to the syndrome, similar to 'El Niño' and 'the Southern Oscillation.' I think the idea was good."[10] After considerable consultation, we ended up adopting the name *unaami*, a Yup'ik word for "tomorrow." It stemmed from recognition that the observed changes taking place were making it more difficult for people in the North to predict what the future would bring. The stated core aim of SEARCH was to understand unaami. We had a mission. We started writing.

CLOSING THE CENTURY

The period spanning the birth of SEARCH in late 1998 to the end of the century was eventful on many fronts. Evidence grew that, at least in some areas, permafrost was continuing to warm. The evidence for hydrographic changes in the Arctic Ocean grew. Despite the high September sea-ice extent of September 1996, the overall downward trend in ice extent for the Arctic as a whole persisted and firmed up.[11] In other words, unaami was becoming more prominent. September 1998 had a new record low ice extent in the Beaufort and Chukchi Seas. Meanwhile, Drew Rothrock of the University of Washington, along with Gary Maykut and Y. Yu, had been looking at submarine sonar records of ice draft.[12] Ice

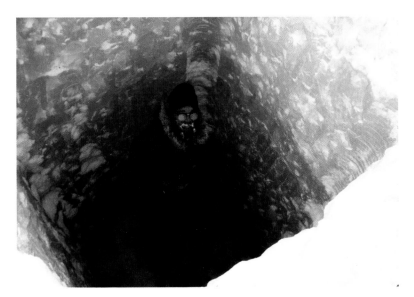

PLATE 1. This is what sea ice used to look like. This photo shows American scientist Roger Andersen during the spring of 1981 at the Office of Naval Research Fram III camp north of Fram Strait. Roger, along with Alan Gill, spent two days with chisels, shovels, a ladder, and buckets digging through 12 feet of ice to create this hole, which would allow scientists to lower instruments through the ice and make measurements of the ocean water underneath. The photograph was by either Tom Manley or Jay Ardai, both from the Lamont Doherty Geological Observatory.

PLATE 2. Canadian glaciologist Roy (Fritz) Koerner on an outlet glacier draining the Devon Island Ice Cap, during April 1983. Fritz was impervious to cold. Photograph by the author.

PLATE 3. Dropping off fuel for the helicopter, on the sea ice near Resolute Bay, Northwest Territories (now Nunavut), Canada, spring 1991, with cold-tolerant Canadian graduate students looking on. Photograph by the author.

PLATE 4. Weathering an impressive spring blizzard, May 1992, on the sea ice near Resolute Bay. Because of the eruption of Mount Pinatubo in June of 1991, which injected sulfate aerosols high into the atmosphere, summer never really came to the Canadian High Arctic in 1992. Few people were thinking about Arctic warming. Photograph by the author.

PLATE 5. The author, May or June 1982, downloading data at station "Zebra," located roughly in the middle of the larger of the St. Patrick Bay ice caps. While attempting to recharge the thermal printer, we inadvertently fried it; from then on, all data had to be written down by hand. Photograph by Ray Bradley. Courtesy of the author.

PLATE 6. Camp, on the edge of the larger of the St. Patrick Bay ice caps, June 1982, with a rare convective-type cloud passing nearby. There was enough electrical activity in the area to scramble one of the data recorders. It never occurred to us that, this far north, we would need to ground the weather instruments. Photograph by Ray Bradley. Courtesy of the author.

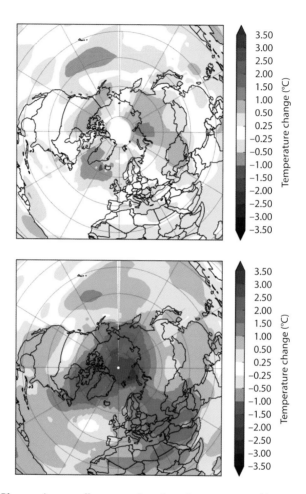

PLATE 7. Changes in annually averaged surface air temperature (degrees C) over the 30-year period 1957 to 1986 (top) and for the 30-year period 1987 to 2016 (bottom). In the earlier period, when natural climate variability was the dominant feature (in particular, that associated with the North Atlantic Oscillation and Arctic Oscillation), some parts of the Arctic were warming while others were cooling. In the later period, it is warming pretty much everywhere, and the Arctic amplification is obvious. Created by Alenxander Crawford. Courtesy of the author.

5 September 1980

▨ Median ice edge 1981–2010

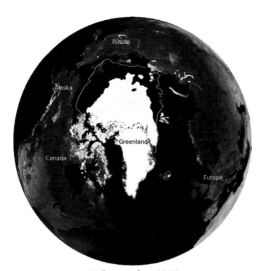

17 September 2012

▨ Median ice edge 1981–2010

PLATE 8. Arctic sea ice at its minimum summer extent, on September 5, 1980 (top), near the beginning of the period of satellite coverage, and on September 17, 2012 (bottom), the record low year as of this book. The September minimum extent in 2012 was only about half of that recorded in 1980. The orange line shows the average location of the ice edge from 1981 to 2010. Courtesy of NASA.

PLATE 9. McCall glacier, in the Brooks Range of Alaska, photographed in 1958 by Austin Post (top) and in 2004 by Matt Nolan (bottom, courtesy of the National Snow and Ice Data Center, University of Colorado, Boulder). The shrinkage of the glacier between the two photograph panoramas is striking.

PLATE 10. The terminus of the McCall glacier, in the Brooks Range of Alaska, photographed in 1958 by Austin Post (top) and in 2003 by Matt Nolan (bottom, courtesy of the National Snow and Ice Data Center, University of Colorado, Boulder). The retreat of the glacier terminus between the two photographs is striking.

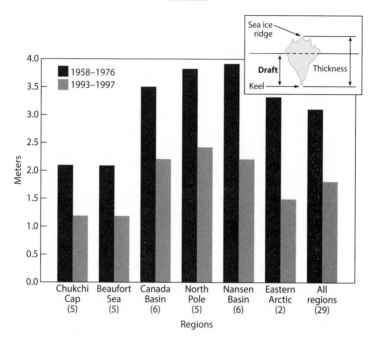

FIGURE 20. Decrease in Arctic Ocean sea-ice draft between the periods 1958–1976 and 1993–1997. Courtesy National Snow and Ice Data Center (NSIDC).

draft is the part of the sea ice that lies below the surface (roughly 90%, but it varies a bit) and that is measured by upward-looking sonar. The roughly 10% that sticks above the water surface is called the freeboard. When they compared ice draft data collected between 1993 and 1997 with earlier records for the period 1958–1976, they found that the average ice draft between the two periods had decreased by 1.3 meters in most of the deep-water regions of the Arctic Ocean (submarines understandably avoid shallow areas) (fig. 20). This was the first really solid piece of evidence that the observed decline in sea-ice extent was attended by a thinning of the ice cover. In other words, ice volume was shrinking.

Other field programs were taking off, such as the NSF ARCSS program called Arctic Transitions in the Land-Atmosphere System (ATLAS) to address transfers of carbon dioxide, methane, water, energy, and nutrients between the land surface and the atmosphere. I was able to participate in both the 1999 and 2000 ATLAS field seasons. Activities focused on sites around the village of Council, Alaska (summer population of perhaps 20, winter population about zero), roughly 60 miles from Nome, on the banks of the Niukluk River. Council had seen its heyday in 1897 and 1898, when gold was discovered in Ophir Creek, a small tributary of the Niukluk. Perhaps 15,000 people once lived there, but they left when larger deposits of gold were discovered around Nome in 1900. There is a lot of history lying around Council, including an immense dredge and Bucyrus steam shovels. There are also very large brown bears and moose walking around, and one is obliged to exercise caution when hiking about.

The terrain is a mix of tundra and shrubs in the lowlands and boreal forest in the higher elevations, where it is better drained. Selection of the Council area for ATLAS reflected the thinking that this would be representative of what tundra environments, such as those presently characterizing the North Slope of Alaska, would transition into with anticipated climate warming. One of my roles was to launch radiosondes from tundra and forest locations (the latter from a small clearing) to see how the different environments were affecting vertical profiles of temperature, humidity, and winds. I also assisted the botanists in assessments of biomass.

We worked hard and learned a lot. Apart from a few over-privileged students from Middlebury College engaged in tree coring (they became known as the "tree people"), everyone got along, despite the overcrowded, damp, and highly unsanitary conditions. Cooking for 20–25 people at a time could be a challenge, but there was never a shortage of moose meat in various forms brought to us by the local villagers. This included moose heart and moose nose. Sometimes it could be moose for breakfast (moose patties), moose for lunch (moose burritos, or "mooseritos"), and roast moose for dinner. On special occasions, we might also have moose for brunch.

Partway through the first season, a young scientist came to camp with her sick infant, which was widely, if perhaps unfairly, blamed for a nasty stomach virus that quickly passed through camp. We eventually all recovered, including the hapless tot. Meanwhile, the helicopter pilot, who we'll call Jim Spring for his own protection, kept us entertained with ribald humor, poker skills, and an impressive off-duty capacity for Wild Turkey. Its antiseptic qualities may be why he escaped the stomach illness. He was very careful about his drinking. The rule, as he claimed, was that a pilot must not consume alcohol within 12 hours of having to fly. He would look at his watch, and take the evening's last shot of Wild Turkey literally seconds before curfew.

While personally fine with the term unaami, I was becoming a little concerned that the SEARCH Science Plan was becoming framed too heavily around the AO. I'd been doing a fair bit of my own work focused on understanding variability in patterns of atmospheric circulation in the Arctic, and while imprints of the AO and NAO were certainly clear, it was obvious that there was a lot more going on. The AO explained approximately 25% of the variability in winter circulation at sea level from the Northern Hemisphere, and while this is pretty good, it meant that 75% was other stuff. One could plot up average sea-level pressure fields from two winters with essentially the same AO index, and they could look very different.

Other scientists, like Clara Deser of NCAR, fretted about the argument for the AO being more "fundamental" than the NAO.[13] As Clara put it, "When Thompson and Wallace introduced the concept of an Arctic Oscillation as a mode of circulation variability more fundamental than the well-known North Atlantic Oscillation, I questioned the strength of the linkage between the North Pacific center of action and the centers of action over the Arctic and Atlantic that are common to both the AO and NAO. Simple analysis of the data showed that the Pacific center was not significantly correlated with the others, and that this was the case over a wide range of timescales. Thus, the AO did not appear to represent a single dynamical pattern. Later studies by others corroborated this result. Nonetheless, the AO is appealing in its simplicity and because it mirrors the Antarctic Oscillation in the Southern Hemisphere. But one must keep in mind that the Pacific center of action in the AO is only very weakly associated with the rest of the pattern."[14]

Amanda Lynch, an atmospheric dynamics scientist now at Brown University and one of the leaders of the ATLAS effort, saw it similarly. "The Antarctic region has a high degree of what we call 'zonal symmetry'— that is, as you travel a circle around the pole, there won't be much variation. It will be all ocean, all high ice plateau, depending on how far south you are. The Arctic is much more complex; as you travel a circle at any northern latitude, you might see floating sea ice, then rocky mountain ranges; open ocean, then forests; vast tundra plains, then steep ice sheets. For that reason, the tendency for the spinning earth and atmosphere to give rise to a simple annular signal like the Antarctic Oscillation is continually disrupted by the surface variation."[15]

So if the Pacific and Arctic centers were not correlated, but the Arctic and Atlantic centers were, then this says that the AO is essentially just the NAO. In this regard, looking back at figure 18, the focal points of the Atlantic centers of action in the AO correspond almost precisely to the NAO centers of action—the Icelandic Low and Azores High. Regarding the argument that if the planet was flat and featureless (no distorting influences), the AO would look annular, the reality is that the earth is not flat and featureless, so you can't really say that the AO is inherently annular. Despite such concerns, the AO stuck as a leading paradigm in Arctic climate research and a centerpiece of unaami.

Meanwhile, I led an effort to try and synthesize all of the available evidence of recent change in the Arctic into one paper.[16] It was inspired in part by involvement in SEARCH but largely reflected my own need to try and make sense of things. Among the coauthors were John Walsh, Jamie Morison, and Roger Barry (my dissertation adviser back in the mid and late 1980s), along with experts on permafrost (Tom Osterkamp, Vladimir Romanovsky), terrestrial ecology (Terry Chapin, Walt Oechel), and glaciers (Mark Dyurgerov). We looked at changes in the Arctic Ocean, temperature, atmospheric circulation, glacier mass balance, forest fire frequency, and photosynthetic activity over land and made comparisons with projections of Arctic change from the climate models of the time. After numerous revisions that I thought would never end (as I recall, peer reviewer #2 was a real pain), the paper was published in April 2000 under the title "Observational Evidence of Recent Change in the Northern High Latitude Environment."

The second paragraph of the abstract (the summary of the paper) capsulated our view: "Taken together, these results paint a reasonably coherent picture of change, but their interpretation as signals of enhanced greenhouse warming is open to debate. Many of the environmental records are either short, are of uncertain quality, or provide limited spatial coverage. The recent high-latitude warming is also no larger than the interdecadal temperature range during this century. Nevertheless, the general patterns of change broadly agree with model predictions. Roughly half of the pronounced recent rise in Northern Hemisphere winter temperatures reflects shifts in atmospheric circulation. However, such changes are not inconsistent with anthropogenic forcing and include generally positive phases of the North Atlantic and Arctic Oscillations and extratropical responses to the El Niño Southern Oscillation. An anthropogenic effect is also suggested from interpretation of the paleoclimate record, which indicates that the 20th century Arctic is the warmest of the past 400 years."

Simply put, despite the growing evidence for a role of greenhouse warming in both global and Arctic change, we were still not convinced. Again, it wasn't a question of whether a human influence on Arctic climate was to emerge; it was still a question of whether it had. And to my mind, the answer was still a very resounding "maybe," for despite all of the changes that were being seen in the Arctic, a lot of unaami still looked like natural climate variability. But a turning point was near.

5

EPIPHANY

The new century dawned. Arctic changes became ever more obvious and now clearly included the Greenland ice sheet, ice caps, and glaciers. Even the landscape was changing, with areas of treeless, windswept tundra being taken over by shrubs. By now, many Arctic scientists were convinced that we'd moved beyond just natural variability. But I was not the only one still sitting on the fence. The expected Arctic amplification still wasn't especially prominent. And, from my point of view, as it became ever more apparent that so much of what was happening was linked to the AO, it became less apparent why one needed to invoke the specter of rising greenhouse gas levels to explain unaami. Then, at the height of the AO mania, something remarkable happened. The AO, and its little sister the NAO, began regressing from their high positive phases. But the Arctic kept warming and the sea-ice cover kept shrinking. It was sometime in 2003 that I saw the light. Unaami was more than just a natural climate cycle. It was us.

THE THIRD IPCC ASSESSMENT

Contrasting with the guarded statements expressed in our paper published in 2000, "Observational Evidence of Recent Change in the Northern High Latitude Environment," the Third IPCC Assessment Report (TAR) of 2001, expressed high confidence that for the globe viewed as a whole, the human fingerprint of climate change was clearly visible.[1] The data firmly pointed to an increase in global average temperature of about 0.6°C over the 20th century. The 1990s ended up as the warmest decade in the instrumental record. Analysis of paleoclimate data indicated that the temperature increase during the 20th century was probably the largest of any century during the past 1000 years. The IPCC Summary for Policymakers stated, "There is new and stronger evidence that most of the warming observed over the last 50 years is attributable to human activities." However, the report also stated that some aspects of climate had not changed. For example, Antarctic sea ice had remained fairly stable.

The role of aerosols got a lot of attention. It was better understood that the short-lived aerosols injected into the troposphere from fossil fuel burning mostly cool the atmosphere by absorbing and scattering solar radiation before it can reach the surface. Short-lived means that the residence time of tropospheric aerosols is only a few weeks, because they quickly get washed out. Still, as quickly as they are washed out, they are replenished. By contrast, explosive volcanic eruptions, like Pinatubo, inject cooling sulfate aerosols high into the stratosphere, where they last much longer. Hence the effects of fossil fuel burning have competing effects; the carbon dioxide leads to warming, but the aerosols mask some of that warming. We also began to realize that aerosols have important indirect effects in that they can influence the albedo and lifetime of clouds, which greatly affect the energy balance at the surface.

The early 2000s saw continued advances in computing power, enabling more detailed mathematical representation of physical processes in climate models. The IPCC TAR expressed greater confidence in the ability of climate models to project future conditions, but the projections ranged widely. For example, it was projected that the average surface temperature for the globe would increase anywhere from 1.4°C to 5.8°C over the period 1990–2100, with a rise in sea level of anywhere from 0.1 to 0.9 meters over the same period. This reflected a combination of uncertainty about future rates of greenhouse gas emission and the fact that climate models still had considerable shortcomings, associated with things like the physical treatment of clouds and their interaction with radiation and aerosols.

AO MANIA

In 2002, as part of his PhD work, Ignatius Rigor at the University of Washington, in Seattle, teamed up with

Mike Wallace and Roger Colony to take a careful look at links between the behavior of the Arctic Oscillation—or the AO—and sea-ice extent.[2] It was common knowledge that the increasingly firm downward trend in total September sea-ice extent was dominated by ice losses along the Siberian and Alaskan coasts. What wasn't so clear was why. Ignatius showed that as the winter AO index climbed to positive values in the 1990s, there was a shift toward a more cyclonic (counterclockwise) pattern of sea-ice motion, consistent with a shift toward a more cyclonic pattern of sea-level pressure and surface winds, such as had already been shown by John Walsh a couple of years before the AO had been formally discovered. The change in the ice motion tended to pull more ice away from the shores of the Siberian and Alaskan coasts and also promoted more ice fracturing. As the ice was pulled away from the shore, it left open water behind. While the open water areas refroze, the overall effect was to produce more young ice, meaning that, come spring, the ice along the coastal areas progressively thinned. Thinner ice is more apt to melt during summer, which provided an explanation for the large summer ice losses that started to be observed along these shores in the 1990s. In a follow-up study,[3] Ignatius and Mike found that atmospheric circulation patterns associated with positive winter AO conditions, particularly dominant over the period 1989–1995, had also decreased the areal extent of old thick sea ice, mostly by promoting its transport out of the Arctic Ocean through the Fram Strait.

This left the Arctic Ocean with an anomalous coverage of younger and thinner ice. Over the period of several years, this thin ice then circulated back into Alaskan coastal waters via the Beaufort Gyre circulation, where extensive summer ice loss was observed.

This was a really nice piece of work that explained a lot. But it also disturbed me, for it further reinforced the notion that, despite what the TAR had concluded for the globe as a whole, it still wasn't necessary to invoke the Satanic gases (as some were calling them) to explain the changes being observed in the Arctic. That is, unless the trend in the AO was itself some sort of greenhouse warming signal, which was still in the category of the unknowns.

Ignatius's work got a lot of people talking. And, for better or worse, it added momentum to the already well-populated AO bandwagon. As discussed in the last chapter, ever since Dave Thompson and Mike Wallace had published their first paper on the AO in 1998, study after study appeared linking the AO to everything from temperature to Arctic precipitation to river runoff and (with Ignatius's work) sea ice. Jokes were running around that, to get anything published, or to get a proposal funded, one needed to demonstrate that, whatever they were studying, there was a link with the AO. It had arguably become more than just a bandwagon. It had become AO mania.

LAND ICE

One group that didn't seem so much caught up in the AO were the glaciologists studying the Greenland ice sheet. The Greenland ice sheet is one of our planet's two ice sheets; the other is the Antarctic ice sheet. The Greenland ice sheet is much smaller than the Antarctic ice sheet, but it still contains about 2.9 million cubic kilometers of ice, which represents a little more than 7 meters of global sea level (fig. 21). As such, the response of the Greenland ice sheet to a warming climate was getting attention.

In July 2000, Bill Krabill of the NASA Goddard Space Flight Center's Wallops Flight Facility led a team with the mission of estimating the mass balance of the Greenland ice sheet. They used aircraft laser altimeter surveys over northern Greenland from 1993/1994 and 1998/1999, along with other data from the southern part of the ice sheet.[4] Altimeters measure the surface elevation; by comparing elevation data from different years, it is possible to determine the change in elevation, which is related to the change in mass balance. From this sparse data, Krabill's team concluded that the ice sheet was thickening in its higher elevations, but thinning in its lower elevations, the sum yielding a mass loss for the ice sheet as a whole. Hence, the Greenland ice sheet was contributing to sea-level rise.

Every spring and summer, the lower, warmer elevations of the Greenland ice sheet undergo surface melt.

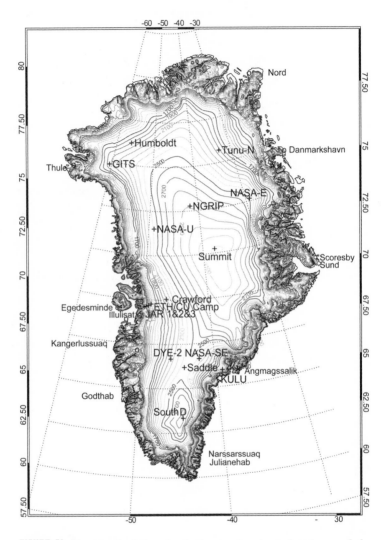

FIGURE 21. The island of Greenland, showing ice sheet elevations and the locations of automated weather stations. Source: Box, J.E., and K. Steffen, K. (2001), "Sublimation on the Greenland Ice Sheet from Automated Weather Station Observations," *Journal of Geophysical Research*, 106(D24), 33,965- 33,981.

How much depends on elevation, latitude, and the energy balance at the snow surface; the latter depends on temperature and the prevailing weather conditions. This meltwater can percolate into deeper snow layers and then refreeze, or in lower elevations, it can form streams and discharge into the ocean. While the change in meltwater runoff is most directly relevant to changing sea level (iceberg discharge is the other major way that the ice sheet can shed mass), simply knowing the total areal extent of melt says a lot about how the ice sheet is changing. A greater areal extent of melt in a given summer generally implies more surface runoff.

In 2001, Waleed Abdalati, then at NASA, and Konrad Steffen, then at the University of Colorado Boulder, presented an analysis of surface melt extent spanning the period 1979–1999.[5] They basically asked: What percent of the ice sheet undergoes melt each summer, what are the spatial patterns, and has the areal extent of melt changed? It had already been shown that the presence of surface melt could be detected using satellite passive microwave data. This is the same satellite data used to monitor sea-ice extent and to map snow extent over land, just applied differently. Over the 21-year record, Abdalati and Steffen documented an upward trend in melt extent over the ice sheet driven mostly by changing conditions over its western part. It was a small trend, and a very noisy one at that—there was a lot of variability in melt extent from year to year—including a cold signal from the Mount Pinatubo eruption in 1991—but a trend nonetheless, adding to the evidence that the mass balance of Greenland had turned negative.

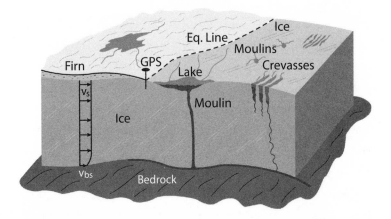

FIGURE 22. Schematic of the factors involved in the Zwally Effect. From Zwally, J., W. Abdalati, T. Herring, et al. (2002), "Surface melt-induced acceleration of Greenland Ice-Sheet flow," *Science*, 297, 218–222. Reprinted by permission of AAAS.

The very next year, in July 2002, NASA glaciologist H. Jay Zwally published an influential study titled "Surface Melt-Induced Acceleration of Greenland Ice Sheet Flow."[6] He showed that at the equilibrium zone (the elevation where winter mass gains are equal to summer losses) on Greenland's west central ice sheet, the ice sheet flow accelerates in summer, nearly coinciding with the duration of summer melting, then slows after the melt is over. Zwally concluded that during summer, there is a rapid surge of meltwater to the base of the ice sheet, reducing the friction at the interface between the ice and the bedrock, so that the ice literally slides more freely along the bedrock surface (fig. 22). The sudden drainage of surface melt ponds through moulins was viewed as key. A moulin is a roughly circular, vertical to nearly vertical, well-like shaft within a glacier or ice sheet through which water drains from the surface to the base. Extension of the process, which came to be known as the Zwally Effect, is as follows: as the climate warms, more meltwater is produced, which enhances the sliding at the base of the ice, which then increases iceberg calving rates to the ocean from the big glaciers that drain the ice sheet. And, as the climate warms, the area where melt ponds develop and then quickly drain through moulins climbs in elevation, so the enhanced sliding also occurs at higher elevations. Hence, the ice sheet could have accelerating mass loss both by more direct runoff of meltwater and through enhancement of the Zwally Effect. While subsequent work showed that the more important factor leading to sustained acceleration is the thinning of the calving tongues of the glaciers, which reduces what is known as backpressure, Zwally's paper was an important step in understanding how the ice sheet was changing.

As described by Waleed Abdalati, now director of the University of Colorado Boulder's Cooperative Institute for Research in Environmental Sciences (and my boss), events that followed took many scientists by surprise. "Prior to 2000, we could not convincingly say whether the Greenland ice sheet was growing or shrinking, hence raising sea level or lowering it. In 2000, we got our answer [Krabill's study], but we were unaware of what was about to unfold. Beginning in 2003, the calving front of one of the world's fastest-moving glaciers, the Jakobshavn Ice Stream on Greenland's west coast, began a fairly rapid retreat of about a kilometer per year, as melting in this area increased. Subsequently, the ice stream nearly doubled its speed from 7 kilometers per year to 14 kilometers per year, greatly increasing the amount of ice loss from Greenland through this river of ice. In the years that followed, similar behavior was observed at numerous outlet glaciers around the perimeter of Greenland. This increased ice discharge, coupled with increasing surface melt rates, led to a significant increase in the ice sheet imbalance and sea-level contributions."[7]

More information came in about the shrinkage of Arctic glaciers and ice caps. Based on limited data, in 1997 Mark Dyurgerov and Mark Meir of the University of Colorado Boulder were able to document an overall recession of ice caps and glaciers.[8] Five years later, in 2002, a team led by Anthony Arendt of the University of Alaska Fairbanks, completed a major effort to assess changes in the mass balance of Alaskan glaciers.[9] Their conclusions raised eyebrows—between the mid-1950s and mid-1990s, the Alaska glaciers seemed to have thinned by about half a meter per year on average, and the thinning rate was increasing with time, with the mass loss apparently contributing more strongly to sea-level rise than that of the Greenland ice sheet.

Another component of the terrestrial cryosphere that scientists were keeping an eye on was permafrost. As summarized in my 2000 paper, changes in permafrost temperature appeared to be rather spotty, but with an overall pattern of warming. In 2002, Vladimir Romanovsky, the permafrost guru at the University of Alaska

Fairbanks, wrote an assessment of permafrost records as indicators of climate change.[10] As part of this, changes in permafrost temperature were summarized for Alaska, Russia, and Canada. While the record lengths varied between the individual studies that Vladimir drew from, the summary pointed to warming in all areas except northern Quebec. It was recognized that causes for rising permafrost temperature were not always cut-and-dried; in some areas, warming was apparently as much due to changes in winter snow cover as to changes in air temperature. Snow is important because it is a very effective insulator. If the snow is thin, the ground can cool quickly in winter by conducting heat out of the soil and to the surface. A deep winter snow cover limits the conduction, and the ground stays warmer as a result.

SHRUBBIFICATION

Arctic land is mostly tundra: it's too cold for trees to grow, and the vegetation is largely limited to dwarf shrubs, sedges and grasses, and mosses and lichens. In 2001, Matthew Sturm, Charles Racine, and Kenneth Tape documented increased shrub growth, indicative of a warming climate.[11] Back in 1948, as part of oil exploration activities, thousands of low-altitude oblique photographs were taken at locations across Alaska between the Brooks Range and the coast. In the years 1999 and 2000, Sturm, Racine, and Tape took photographs at many of the same locations, allowing them to compare the photographs for changes in the coverage of several shrub species. In numerous photograph pairs, they documented substantial increases in the height and the diameter of shrubs, along with in-filling of shrubs in areas that in 1948 were covered by tundra vegetation.

On a much larger scale, Liming Zhou and colleagues found that for the region between 40 and 70 degrees north (of which at least part is Arctic), about 61% of the total vegetated area in Eurasia experienced a persistent increase in satellite-derived NDVI over the period 1981 through 1999.[12] Recall from chapter 2 that the Normalized Difference Vegetation Index (NDVI) is an index of photosynthetic activity based on the way in which green plants reflect light in different wavelengths. The pattern of change over North America for the same time period was more sketchy. NDVI decreases were found over parts of Alaska, boreal Canada, and northeastern Eurasia; such decreases relate to drought conditions, in which warming occurs without a concurrent increase in precipitation. While one can't make a fair comparison between what Sturm and his colleagues found from photographs of fairly small areas and results from a large-scale NDVI analysis, Zhou's results nevertheless indicated that things were changing, and the clear increase in NDVI over Eurasia seemed to agree with the recent pattern of stronger warming over that continent relative to North America.

TESTING PROJECTIONS

As mentioned earlier, the growth in computing power had led to a new generation of climate models with more complete representations of climate processes. Armed with these models, scientists started taking a closer look at how the Arctic was likely to evolve through the 21st century. In 2003, Marika Holland of NCAR partnered with Cecilia Bitz of the University of Washington to look at the anticipated evolution of Arctic amplification. Recall that Arctic amplification refers to the long-held expectation that as the planet warms, the Arctic will warm the most. The different processes giving rise to Arctic amplification were discussed back in chapter 2. Marika and Cecilia found that when the carbon dioxide concentration was doubled, depending on the model, the Arctic warming exceeded the warming in lower latitudes by a factor of two to four.[13] On the one hand, this big range in projected amplification manifested some of the ongoing challenges of modeling the many physical processes at work and their interactions. On the other hand, the fact that all of the models agreed on the main point—that Arctic amplification was going to emerge—was telling. But was there any evidence from direct observations that Arctic amplification was already here?

We knew that the Arctic as a whole had experienced strong warming from 1920 to 1940, cooling until about 1970 (the cooling period noted by Ray Bradley in his 1972 paper, see chapter 1), then a renewed post-1970 warming. However, the 1920–1940 warming seemed to be just as large as the post-1970 warming. Skeptics (or optimists perhaps) suggested that the recent strong warming was nothing special, especially because it was partly compensated for by cooling over northeastern North America that was quite consistent with the change in the phase of the AO and NAO.

Igor Polyakov and colleagues from the University of Alaska Fairbanks took a rigorous look at Arctic amplification using a data set of annual temperatures averaged for the Northern Hemisphere, which they compared to a separate time series for the Arctic.[14] For the more recent years, the Arctic data set included data from the Russian North Pole camps and drifting buoys. They first computed Northern Hemisphere and Arctic trends for the most recent 17 years of the respective records (1985–2001), then computed a new trend by extending each record back by one year (1984–2001), then back another year (1983–2001), and so on, all the way back to the full record length of 1875–2001. The idea was to see how the Northern Hemisphere and Arctic trends compared when looking at successively longer record lengths. While it is legitimate to question the quality of the temperature time series for the early years when the station records are very sparse, the results were intriguing. For the period 1985–2001, the Arctic temperature trend of an increase of 0.6°C per decade was twice the corresponding Northern Hemisphere value. This argued that the anticipated Arctic amplification based on the climate models, albeit having clear links with the AO and NAO, had become reality. But as the record was extended back, the Arctic amplification became smaller. When the records were extended back 60 to 80 years before 2001, the trend analysis yielded a small Arctic cooling, compared to a small warming for the Northern Hemisphere. Looking back further, there was little difference between the Arctic and Northern Hemisphere trends. Igor maintained that when looking at trends, one ought to look at the longest record possible, and when we did so, the support for Arctic amplification fell apart.

The counter argument is that Arctic amplification is expected to be something that is just starting to emerge, and as such it's entirely appropriate to contrast the trend over the most recent period (e.g., 1985–2001) with trends computed over a longer period. Several efforts followed Igor's. Ola Johannessen and Martin Miles of the Nansen Center in Norway, working with Jim Overland of NOAA, tried to enhance Igor's temperature record using data from the European Centre for Medium range Weather Forecasting (ECMWF) weather model to get better coverage over the Arctic Ocean, where the Arctic amplification signal was expected to be especially strong.[15] Their analysis showed that the warming in recent decades, the warming in the earlier part of the 20th century, and the cooling between these two periods, were all more pronounced in the northern high latitudes compared to the Northern Hemisphere as a whole, despite discrepancies in trend calculations from different data sources due to the types of data and the record lengths examined. Therefore, Arctic amplification meant a bigger change in the Arctic for cooling periods as well as warming periods.

Furthermore, Ola Johannessen's analysis, by including coverage over the Arctic Ocean, provided evidence that recent warming in the Arctic was fundamentally different than the earlier-20th-century warming. The earlier warming event was most pronounced in high latitudes, pointing to some sort of internal natural variability in climate. By contrast, the more recent warming got bigger poleward of the Arctic coast, but was also clearly part of a global warming signal encompassing all latitudes. This struck me as important, for it seemed to be saying that whatever the AO and NAO were doing, the effects were just superimposed on a more general recent warming showing Arctic amplification—which is what the climate models were projecting. It was a definite "aha" moment.

But that "aha" moment was tempered by other findings. Along with Arctic amplification, projections from most climate models pointed to increased precipitation over northern high latitudes. There is always a transport of water vapor from the lower latitudes into the higher latitudes via the atmospheric circulation. This is because much of the planet's evaporation occurs in the lower-latitude oceans. As the moist air moves north, it cools and condenses, and precipitation ensues. The basis of the model projections is that in a warmer climate, the warmer atmosphere can carry more water vapor, and as the moister air cools and condenses on its way north, more precipitation is possible. In general, precipitation over the higher latitudes exceeds evaporation (there is positive "net precipitation"), which is what feeds the big rivers that empty into the Arctic Ocean that play such a dominant role in maintaining the Arctic Ocean's fairly fresh surface layer. While evaporation will also increase in a warmer world, the models were pretty much universal in projecting that this is outweighed by the increase in precipitation, meaning increased river discharge to the Arctic Ocean.

In 2002, Bruce Peterson and colleagues from the Woods Hole Marine Biological Laboratory reported that annual water discharge aggregated for the six largest Eurasian rivers draining into the Arctic Ocean had indeed increased over the period 1936–1999.[16] Now, this is not really a tremendous change—the 7% increase amounts to about 128 cubic kilometers per year more discharge at the end of the record compared to when routine measurements began. This is small compared to the 3200 cubic kilometers per year of total annual discharge for the Arctic Ocean from all rivers. Because of shorter data records, Bruce did not examine what was going on over the North American side of the Arctic. Nevertheless, his findings caused a stir. Changes had been observed in the ocean. Changes had been observed in air temperature and in patterns of atmospheric circulation. Now it seemed that the hydrologic cycle was getting into the act. At about the same time, Daqing Yang, then at the

University of Alaska Fairbanks, showed that there had also been a shift toward more discharge in May and less in June, consistent with warming that would lead to earlier snowmelt.

What was driving the upward trend in annual river discharge? At first glance, it seemed consistent with what the climate models were projecting, and the observed pattern of changing discharge broadly followed the pattern of globally averaged air temperature. But it also broadly followed the NAO index (hence also the AO index). This brought the discussion right back to the recurring conundrum. While the models were projecting increased net precipitation and river discharge as a consequence of warming, the changes that were being observed seemed to have a more obvious link with changes in atmospheric circulation, which might (or might not) simply be a reflection of inherent natural variability in the climate system. We were, of course, thinking about possible anthropogenic influences on the behavior of the NAO and AO, but there was no smoking gun. And Max Holmes, one of the coauthors on the study, recalls some discomfort over how the AO/NAO paradigm seemed to be dominating the conversation. "While we did not think that the AO/NAO was the driver of the observed discharge change, we felt that we'd never get our paper published without exploring that link."[17]

To further complicate matters, the station records gave no clear signal of an increase in Arctic precipitation.

Maybe the precipitation-monitoring network was too sparse and the measurement errors were too high. There are also very few monitoring sites in the higher-elevation mountainous areas where much of the precipitation falls. Gauges also have a big precipitation under-catch problem in the Arctic due to the difficulty in capturing blowing snow. It didn't help that different countries used different types of gauges. The river discharge increases might also be related to something else, such as thawing permafrost, or changes in forest fire frequency altering the hydrology by reducing evaporation. So, as with a lot of the changes that had been observed, trying to figure out the cause only raised more questions.

GETTING OFF THE FENCE

The IPCC TAR had made its case that, from a global perspective, the human fingerprint of climate change had emerged. But in the Arctic, the region where the changes were supposed to be most apparent, the picture was still muddy, at least to me. Yes, sea ice was declining rather sharply at the end of the melt season in September (despite the record high ice extent in 1996), but based on the work of Ignatius Rigor and Mike Wallace, a lot of the decline could be related to the behavior of the AO. The AO and NAO also had a lot to do with patterns of recent temperature change, although the Johannessen paper was cause for thought. While permafrost was warming, questions remained regarding how much of this was due to a warming atmosphere as opposed to changes in snow depth, and changes in snow depth might also simply result from shifts in atmospheric circulation. Annual discharge from the Eurasian rivers was on the rise, but yet again, this seemed to be linked to the AO and NAO. Then Tim Boyd led a new study showing that the cold halocline had partly recovered; the retreat that Mike Steele and Tim had documented in 1998 that had caused so much commotion in the oceanographic community seems have been a temporary thing.[18] Ah, complexity.

On the other hand, results from climate model experiments were coming out showing that without increasing greenhouse gas concentrations, there was no way to get a rise in the global average temperature like what had been observed. I was also struck by a paper published a few years earlier in 1999, led by Konstantin Vinnikov of the University of Maryland, showing trends in Arctic sea-ice extent as simulated from climate model experiments to be much larger than expected from natural climate variations.[19] The probability that the observed September trend for the 1978–1998 period resulted from natural climate variability was less than 2%. For the longer period 1953–1993, the probability dropped to less than 0.1%. A number of assumptions were woven into the fabric of that study, and it was, after all, just a modeling study. Still, their results raised eyebrows.

Don Perovich, chief scientist for SHEBA, was among the scientists just starting to come around: "In the year 2000, I was wallowing in data, but in the happiest sense of the term. I was busy analyzing a wonderful set of observations made during the yearlong drift of Ice Station SHEBA. There was an entire annual cycle of observations over spatial scales from meters to tens of kilometers of albedo, snow properties, melt ponds, ice growth, ice melt, and much more. Analyzing results from 100 ice-mass balance sites, I realized that there was a surprisingly large amount of melting on the bottom of the sea ice in the summer of 1998. What I didn't realize is how much this finding would impact the direction of my research in the years to come."[20]

Others, such as Jen Francis of Rutgers University, had already made up their minds. "My entire career had focused on the Arctic. During the late 1990s, it seemed that a downward trend in the real estate covered by sea ice was emerging. By the early 2000s, there was little doubt that warming owing to increasing greenhouse gases was indeed rearing its ugly head first and was strongest in the Arctic, as had been long expected. My stomach began churning in earnest in August 2003 while attending an NSF-hosted retreat in Big Sky, Montana. Twenty-five researchers from all walks of the Arctic system came together with their own disturbing stories and evidence of change in northern seas, skies, soils, plants, animals, and peoples. The consistent messages were of melt, thaw, disruption, destabilization, warming, moving, weakening, and uncharted trajectories. In research, we often talk of 'ah-hah' moments of discovery, but in this case, it was instead an OMG moment, as we all [many, anyways] came away convinced that the Arctic as we knew it was gone."[21]

Charlie Vörösmarty, now at the City University of New York (the double umlauts in his last name regularly crashed early-generation word processors), a longtime collaborator on Arctic water cycle issues, was also an early adopter.[22] "I wish I could say that there was some specific time and place where the light shone down on me, but I think I was probably more predisposed to accept the human signature than you [referring to me] seemed to be back then. My impression was that the notion of a system in rapid change was swirling around earlier. Part of this came from my exposure in the 1980s and 1990s to the IGBP [International Geosphere-Biosphere Programme] community that was amassing evidence of human influences forcing the planet toward a new state. My adviser back in graduate school, Berrien Moore, also did some early work on the subject beginning in the late 1970s, so I had a front-row seat. While there are certainly non-linearities in climate response, you have to work pretty hard to take on the basic chemistry of Fourier, Tyndall, and Arrhenius on whether or not CO_2 is a greenhouse gas that changes that energy balance."[23]

It was still very confusing to me. I sometimes got the impression that the climate modelers had long ago made the call and were wondering why some observationalists like me were still sitting on the fence. But ever since my work on the little ice caps on Ellesmere Island (unbeknownst to me, they were now already considerably smaller than when I was there in 1982 and 1983), I had been very much an observationalist. I liked real data, however messy it may be. And in line with these roots, I recall telling myself, "If the NAO and AO regress from their high positive phases and the Arctic continues to warm and we continue to lose sea ice, I'll start to be convinced."

And that is exactly what happened.

With all the hoopla about the AO and (to a generally lesser extent) the NAO and links to oceanographic changes, declining sea ice, and regional temperature trends, even by the first couple of years of the 21st century, it was clear that the AO and NAO were starting to regress from their high positive phases back to more neutral conditions. Then the index values basically bounced around between positive and negative. The reality is that they couldn't keep going up. Given the argument in the AO framework that a trend toward an increasingly positive phase means a continued shift in atmosphere mass from the Arctic into lower latitudes, an unabated trend leads to the obviously absurd conclusion that, eventually, the Arctic would become a hard vacuum.

But when the AO and NAO regressed, then started bounding around, the sea ice did not recover. It continued to decline. The Septembers of 2002, 2003, 2004, and

2005 all had extreme sea-ice minima (with respect to the record then available), with 2002 and 2005 both setting new record lows. New work came out in 2004 and 2005 from the University of Washington confirming that the ice cover was thinning.[24] More evidence came out of shifts from tundra to shrub vegetation, and the Arctic continued to warm. While the climate records were still short, it became clear that, beneath the strong climate imprints of the AO and NAO on air temperatures, and, at least at some level, changes in Atlantic inflow and shifting haloclines, there was a background warming starting to emerge.

The weight of evidence turned me. If I had to pin it down, my personal epiphany occurred sometime in 2003. And then I turned hard. Indeed, that very year, I was coauthor on a popular science paper in *Scientific American* provocatively titled "Meltdown in the North."[25]

For other scientists, such as Jim Overland of NOAA, full acceptance took longer. As Jim reminisced "First, in the early mid-1990s, when Jamie Morison and I were developing SEARCH, the hypothesis was that it looked like potential Arctic change, and the key point was that we should start observing multivariate changes [changes in many different variables] to make sure. However, full confirmation for me was in 2008 at a NATO Science Conference. A paper I presented went into a book, and the key line was 'At the end of the 20th and beginning of the 21st century Arctic-wide warming is distinctive from the regional patterns of temperature anomalies that were seen earlier in the 20th century.'"[26]

From one point of view, the trend in the AO that became such a focus of attention—and I think the term AO mania is apt—was a bit of a red herring, for it perhaps distracted the science community from the bigger picture. The broader and, I think, more correct view is that it focused many eyes on the Arctic and forced the science community to challenge itself to understand the interplay between natural variability and forced change. The AO paradigm made us learn a lot about how the Arctic works.

Big events were to follow in the next five years or so. That included storm clouds of the political kind.

6

RUDE AWAKENINGS

Like any human endeavor, science is prone to human frailties. Mistakes can be made in data analysis and interpretation. Scientists are, as we have seen, sometimes prone to jumping on the bandwagon. But it's important to remember that the scientific process tends to be self-correcting. Where human frailties really get in the way is when science gets caught up in politics. Pork-barrel spending led to bad blood in the Arctic research community. Politics and greed are old friends, and when they teamed up, SEARCH stumbled. As public awareness grew regarding the reality of climate change, so did efforts to bury the truth through suppression of data and information, and harassment and intimidation of prominent scientists. Try as they might, those who sought to discredit the science could not bury the events that were to unfold in the Arctic. The record-low September sea-ice extent of 2007 sent shock waves through the science community; we'd never seen anything like this before. The long-awaited Arctic amplification signal linked to sea-ice loss emerged in force. Along came 2012, blowing the old record-low ice extent of 2007 out of the water. A sense of inevitability crept into the tone of the discussion. It was no longer a question of whether the Arctic Ocean would lose its summer sea-ice cover. It was only a question of when the clock would tick to zero.

POLITICS

Academic research, whether in the United States or abroad, is by and large funded by the government—that is, by you and me, the taxpayers. I'm referring here to not-for-profit research; the pharmaceutical industry, for example, does a lot of research, but this is profit-driven. In the non-medical sciences, the key funding agencies are NSF, NASA, and NOAA. NSF is largely focused on basic research. It has parallels in other nations. For example, Canada has the Natural Sciences and Engineering Research Council of Canada (NSERC). The Nordic countries collaborate via NordFosrk. NASA and NOAA are more mission-oriented U.S. agencies, but they also fund basic research. Among its mission responsibilities, NOAA handles weather forecasting. Any weather forecast for the United States that you see on the nightly news or get from the Internet or over the radio ultimately uses output from numerical weather models operated by NOAA. NASA handles earth-orbiting environmental satellites (NOAA has satellites too), exploration of other planets, and the manned space flight program. Environmental monitoring has been the focus of the NASA Earth Observing System and the upcoming Decadal Survey featuring the next generation of NASA's environmental monitoring satellites.

NSF, NOAA, and NASA, like other U.S. agencies, put together annual budgets, which become part of the president's overall draft federal budget, which then goes to Congress. Typically, budgets will pass Congress only after a multitude of changes. The U.S. Senate Appropriations Committee plays a big role; the Constitution requires "appropriations made by law" prior to the expenditure of any money from the Treasury. The chairman of the Appropriations Committee holds considerable power, which, invariably, has been used (or abused) to bring his or her state special projects called earmarks, also known as pork (earmarks were banned in 2011).

Via the power of Senator Ted Stevens, then senator from Alaska and chairman of the Senate Appropriations Committee, NSF became the target of pork-barrel spending. Mike Ledbetter, then the director of the NSF Arctic Climate System Study (ARCSS) program and a driving force behind SEARCH, agreed to talk to me and relate what happened from his point of view.

"First you need to know that the management of OPP [the NSF Office of Polar Programs] never embraced the Arctic side of the OPP house [OPP dealt with both Arctic and Antarctic science]. In a good year, OPP requested the same percentage funding increase for the Arctic as the Antarctic, but that, of course, meant that the huge Antarctic program, which included an enormous logistics budget, got large increases while keeping the Arctic program relatively small. Only after Senator Ted Stevens stepped in to increase the Arctic logistics budget did the Arctic ever get a steep increase in budget,[1] but that came with a price because he also demanded that NSF give—and, yes, I mean give—$5 million annually to the UAF [University of Alaska Fairbanks]–Japanese collaborative program."[2]

The UAF–Japanese collaborative program that Mike refers to was at IARC—the new International Arctic Research Center. Basically, Ted Stevens appropriated the NSF money to match the investment of the Japan Agency for Marine-Earth Science and Technology (JAMSTEC) in IARC. And the $5 million was not a one-off deal; it was to be an annual installment. So, while one might say that this is just politics, and that Ted Stevens should be highly commended for pushing through an increase in the Arctic section OPP budget, the $5 million per year of NSF funding to IARC meant $5 million less to the broader research community that could have been the spark that SEARCH and other projects needed.

"As I was told the story," continued Mike, "Stevens threatened the NSF director with the potential of holding up the entire NSF budget if NSF did not send the $5 million to UAF in the next 30 days. We did not even have a formal proposal from IARC for the $5 million, only an empty jacket that was forwarded to the grants office. This was done in spite of the rule by the NSF Science Board that grants that large had to go through them for approval. The Science Board was told what happened in a closed-door session the next time they met. I seriously doubt there is a written record of that part of the meeting."

The way in which individual scientists or collaborating groups normally get research funding from the NSF (as with other agencies) is through a competitive peer-review process. Writing a strong proposal is a time-consuming and tedious task. The proposal must describe the nature and value of the research, with strict adherence to NSF requirements on length, format, and content. The proposal may be in response to an annual call for proposals due on a certain date each year, or to a targeted research opportunity such as those being developed within ARCSS. The proposal is then sent out for peer review and may also be reviewed by a panel. On the basis of the reviews, the work is either funded or not. It is a deliberate and thoughtful process, and though it has its flaws, just like the anonymous peer-review process for manuscripts, it is the best approach that anyone has come up with so far, and has endured. It can be gut-wrenching, and you have to have a thick skin to deal with bad reviews, but it is basically an open and fair process.

The block funding to IARC was an end run around the established process. International collaborations are great things, and we need more of them, and the JAMSTEC-IARC collaboration led to a lot of good work. The issue was not the intent but rather the process. Not surprisingly, the research community was outraged. A number of proposed efforts focused on a new initiative for long-term Arctic observatories were threatened. This included a project spearheaded by Jamie Morison of the University of Washington that would eventually be called the North Pole Environmental Observatory (NPEO), an automated scientific observatory over the central Arctic Ocean intended to monitor ocean, atmosphere, and sea-ice conditions.

As Jamie remembers it, "A bunch of us were mad about having the rug pulled out from under our proposals and wrote an e-mail to the NSF complaining about it. I was lead author, and the letter was always spoken of as 'Jamie's letter.' IARC held several meetings to try to get volunteers to write up a science plan for IARC [to guide use of the funding], but that didn't get a lot of enthusiastic support."[3]

IARC hired new staff scientists, at least partly supported by the $5 million per year. Feeling some guilt, and certainly under some pressure, IARC for a time tried to become a de facto secondary funding agency, shelling out some of the $5 million to support other research. I wanted nothing to do with what I viewed as tainted money and steered clear. Eventually, though, hard feelings blew over, and under Larry Hinzman, who assumed the IARC directorship in 2006, IARC prospered.

Long-term observatories also took off; Jamie's NPEO got its first funding in 2000.

So much for the purity of the peer-review funding process. And there was much more going on at the time than just the pork-barrel imbroglio, which ended up spelling trouble for SEARCH.

As Mike continued, "When I started to push SEARCH as an ARCSS initiative due to its integrated science nature, a decision was made to expand the initiative to a section-wide initiative rather than an ARCSS-only initiative. With the help of the Polar Research Board, a strong interagency planning effort began. Just about the time I left NSF, it was expanded to an interagency initiative, something like SHEBA but more expansive in scope, and then to an international initiative through IASC [the International Arctic Science Committee].[4] At that point, everyone who could spell 'Arctic' jumped in to try to direct where the initiative went because every agency, non-governmental organization, and international partner saw it as a potential golden egg laid by the United States through NSF. Of course, with that many hands on the steering wheel, the planning escalated, but the funding never showed up. After I left, I was so disgusted with the non-direction of ARCSS/SEARCH/OPP that I quit reading about what was happening, but I definitely got the sense that the initial momentum was waning. It takes strong leadership to make scientists cooperate, especially when interdisciplinary science is concerned. I have my take on whom to blame, but I was not there when SEARCH turned into an excuse to have planning meetings instead of a funded research project with well-organized objectives matched with research funding. I am sure the Arctic NSF management was up against a lot of Antarctic momentum, including the creation of a new ARCSS-like Antarctic program that is bound to have sucked up a chunk of the OPP budget."[5]

No doubt there are those who will disagree with Mike's interpretation of events, and Mike, a forceful personality within NSF, had his share of detractors. But, from what he related, it seems that the old cliché holds true—too many cooks spoil the broth. Mike stepped down from NSF in late 2002 to pursue other interests. For a time, SEARCH, living on its own momentum, stormed on. Following release of the Science Plan in 2001, there was a SEARCH Workshop on Large-Scale Atmosphere Cryosphere Observations, with the aim of identifying the observations needed to track Arctic change. This was followed in 2002 by a SEARCH Human Dimensions Workshop. The U.S. Arctic Research Commission, a U.S. federal agency established by the Arctic Research and Policy Act of 1984, met to discuss the project. A SEARCH Open Science Meeting was held in October 2003 to great fanfare. There was still a feeling in the air that SEARCH was going to blossom into a big international, interdisciplinary, and multiagency endeavor focused on understanding Arctic change—that is, unaami. More than 400 people attended the Seattle meeting;

there was strong international representation, including scientists from Canada, the United Kingdom, Germany, France, Finland, Norway, Sweden, Russia, Japan, South Korea, China, Australia, and other countries. International coordination then grew further with SEARCH for DAMOCLES (S4D; Developing Arctic Modeling and Observing Capabilities for Long-term Environmental Studies is a European consortium of institutions working to develop an Arctic atmosphere-ice-ocean observing system). The JAMSTEC folks at IARC were also playing a prominent role.

But over the next few years, SEARCH slowly lost momentum, and as several scientists described it to me, SEARCH enthusiasm was replaced by "SEARCH fatigue." SEARCH still continues, and it has perhaps found some footing as a mechanism for scientists to get together via cross-cutting working groups to synthesize science so that it can be of value to decision makers (everything from managers of shipping companies to politicians). Still, I agree with Mike Ledbetter that SEARCH never realized its potential. Science can be filled with disappointments.

Around the same time as the IARC fiasco unfolded, there was an increasingly unfriendly attitude toward climate science in the United States on the part of the George W. Bush administration, including harassment and intimidation of scientists. Ray Bradley, my mentor from the University of Massachusetts, who first introduced me to the Arctic, was one of the targets of the political backlash, which inspired his book titled *Global Warming and Political Intimidation: How Politicians Cracked Down on Scientists as the Earth Heated Up*.[6]

As Ray summarized, "Concern over human-induced climate change gathered momentum in the early 2000s following publication of the Third IPCC report. Momentum grew further when Vice President Al Gore made global warming the main topic of his post-governmental career, speaking out forcefully and [a few years later in 2006] publishing his award-winning book and movie *An Inconvenient Truth*. All this attention put pressure on members of Congress to introduce legislation to reduce greenhouse gas emissions. However, the energy industry was not about to let that happen and used their influence with several key members of Congress to attack the science of global warming. Rather than present compelling alternative explanations for the observed warming, they chose to attack the credibility of the scientists involved. Their goal was to sow the seeds of doubt about the veracity of the scientific evidence, a tactic straight out of the playbook of the tobacco industry, which had used the same approach years earlier to delay legislation over tobacco use."[7]

For some scientists, the backlash was limited to puzzling exchanges with government officials who may have just been trying to cover themselves. One that I was personally aware of, because it happened here at the University of Colorado, involved Koni Steffen, who had been doing work on melt of the Greenland ice sheet. As Koni relates, "I received a phone call from Dr. James R. Mahoney, assistant secretary of Commerce for Oceans and Atmosphere and deputy NOAA administrator, special assignment as director of the U.S. Climate Change Science Program, who stated that if I give press briefings about Greenland melt I should contact his office so they would be connected by a conference call—or I should brief them first on what I was going to say to the press. During that time, I was the director of the CIRES, which is supported by NOAA. I countered his suggestion by stating that NOAA support notwithstanding, I was an employee of the University of Colorado. He then said that I had to mention every time I gave a press interview that I was not with NOAA. A few weeks later, I related that telephone conversation to a journalist from the *Washington Post* who was writing a short article. When she called Mahoney to get his opinion on our call, Mahoney said he had no recollection at all of our conversation. Whether Mahoney was pushed by the administration or acted out of his own initiative is unclear."[8]

Ray Bradley, by contrast, was caught right in the middle of a smear campaign, along with his colleagues Michael Mann and Malcolm Hughes. It started with a graph from their work, published in the Third IPCC Assessment Report. The graph, based on paleoclimate reconstructions, showed a slight cooling over the last millennium, followed by abrupt 20th-century warming.[9,10] Climate scientist Jerry Mahlman coined the term "hockey stick" to describe such graphs, which have by now been reproduced in many independent studies—relatively unchanging temperatures up to around 1900 representing the shaft of the hockey stick, then a sharp rise representing the blade of the stick. Although the conclusions of the IPCC would have been the same had the hockey stick graph not been included, thanks to media attention, in the mind of the public the hockey stick became synonymous with the IPCC's conclusions.

As Ray recalls, "We ended up being targeted by Texas Republican Joe Barton, who chaired the House Energy Committee. Although that committee had never held any hearings or shown any previous interest in the topic of global warming, Barton demanded scientific records, e-mails, and financial statements spanning our entire careers, an action that was unprecedented. Fortunately, another Republican, Sherwood Boehlert of New York [chairman of the House Science Committee] saw the attack for what it was—a blatant effort to intimidate scientists—and demanded that Barton back off. Their dispute created so much media attention that the demands were dropped, but the fallout has continued for more than a decade, with lawsuits from right-wing groups demanding access to e-mails still pending, and countersuits claiming defamation still in the courts."[11]

NEW INITIATIVES

But science is resilient. The NSF ARCSS Freshwater Integration (FWI) study, designed to understand the flows of water through the Arctic system, started in 2002. Now the science community had a program that would allow us to determine why river discharge was changing and get a better handle on things like the oceanic flows of water into and out of the Arctic Ocean. Phase II of Shelf Basin Interactions got underway, addressing the transformation and fate of carbon at the edge of the Arctic's continental shelves. In 2004, the Study of the Northern Alaska Coastal System started to investigate vulnerability in the natural, human, and living systems of the Alaska coastal zone and impacts from current and projected environmental change.

The development of these U.S. programs and others that followed was aided by the ARCSS committee, consisting of scientists from various disciplines chosen to serve because of their ability to listen and respond to a broad science community. I was honored to serve on the ARCSS committee for several years, and I also served on the SEARCH Steering Committee. These activities kept me plugged into what was going on.

Charlie Vörösmarty remembers how the synergy with the ARCSS committee helped to launch the highly successful FWI project: "Bruce Peterson [Woods Hole Marine Biological Laboratory], who was on the ARCSS committee, approached me and Larry Hinzman [UAF] about organizing a workshop on designing an integrated study of Arctic hydrology. We received funding from NSF and held the workshop in September 2000 in Santa Barbara at the National Center for Ecological Analysis and Synthesis [known as the NCEAS]. The workshop resulted in the Arctic CHAMP report [Community-wide Hydrologic Analysis and Monitoring Program]. From the get-go, our thinking was interdisciplinary and the participants, from both the U.S. and abroad, who contributed to the report were drawn from the arenas of atmospheric, land surface, and oceanic Arctic sciences and the social sciences. Some were observationalists, some experimentalists, some modelers, and some remote sensing experts. The resulting report was evidently in the right place at the right time, and Mike Ledbetter [then still the ARCSS director] ultimately funded the FWI."[12]

GETTING THE WORD OUT

The year 2004 saw publication of the Arctic Climate Impact Assessment, or ACIA.[13] The ACIA was an international project of the Arctic Council and the IASC, dedicated to evaluating and synthesizing existing understanding of Arctic climate variability and change and their impacts. Remember, as discussed in chapter 2, that the Arctic Council is an intergovernmental forum with members from Canada, Denmark, Finland,

Iceland, Norway, the Russian Federation, Sweden, and the United States. The primary focus of the ACIA was on the physical environment, but it also included a comprehensive assessment of economic impacts and impacts on the peoples of the Arctic. Results of the assessment were released in November 2004. It was a major effort, lavishly illustrated and involving many scientists. It was a great example of international collaboration and ended up being translated into several languages.

The report stated, "The Arctic is extremely vulnerable to observed and projected climate change and its impacts. The Arctic is now experiencing some of the most rapid and severe climate change on earth. Over the next 100 years, climate change is expected to accelerate, contributing to major physical, ecological, social, and economic changes, many of which have already begun. Changes in Arctic climate will also affect the rest of the world through increased global warming and rising sea levels." The key findings that followed didn't leave much room for uncertainty. Big things were already happening, and more was on the way.

The ACIA got a lot of attention, and for some, like Volker Rachold of the Alfred Wegener Institute for Polar and Marine Research (Germany), it was a turning point in their thinking. Volker, who specializes in permafrost, nutrient and carbon cycles, sedimentation processes, and river geochemistry, said, "Already in the early 1990s, when I did my PhD on black shales and orbital cycles during Cretaceous times [back in the age of the dinosaurs, a very warm period in earth history], I was convinced that at some point we would see the impact of greenhouse warming. The analogy between the CO_2-driven Cretaceous greenhouse world and the present situation seemed very obvious to me. At that time, I did not think about the Arctic yet, and the observations were still pretty vague anyway. The turning point was mid-2000s, when the Arctic Climate Impact Assessment was published. I believe that I was not the only one who was convinced by the findings compiled in this assessment."[14]

While the weight of evidence had now fully convinced me that the human imprint on climate change had emerged in the Arctic, I still felt that parts of the ACIA were a bit strident in their presentation and could have better emphasized that natural variability was still a big player. But since I had contributed to the assessment, albeit not as a lead author, I didn't have much of a leg to stand on and had yet to come to grips myself on how to best communicate what was going on in the Arctic.

Indeed, I was sometimes a picture of contradiction, and depending on who I was talking to, my voice changed. In 2004, I gave testimony to the U.S. Senate Committee on Commerce, Science, and Transportation (Sen. John McCain presiding) regarding the Arctic sea-ice cover. McCain was known as a maverick in the Republican Party, and he expressed what seemed to be genuine concern over the lack of action by the U.S. government on climate change. Indeed, he pushed the unsuccessful McCain–Lieberman Climate Stewardship Act of 2003, which would have required power stations to reduce their emissions. And because I fancied having some inkling as to how politicians think (I have since realized that I don't), my testimony didn't have any nuance. I flat out said that sea ice was declining and that the culprit was global warming. This I justified in the spirit of a phrase I'd picked up: "When speaking to the hard of hearing, one should speak a little louder."

What I left out was that imprints of natural variability were also very prominent, something strongly emphasized in a sea-ice paper that I was to publish a year later.[15] We stated, "While the record minimum observed in September 2002 strongly reinforced this downward trend [over the period of satellite observations], extreme minima were again observed in 2003 and 2004. While having three extreme minimum three years in a row is unprecedented in the satellite record, attributing these recent trends and extremes to greenhouse gas loading must be tempered by recognition that the sea-ice cover is variable from year to year in response to wind, temperature and ocean forcings."

It was an honor to be asked to testify to Senator McCain, although I can't see that it had any effect. In later years, McCain largely reverted to toeing the Republican line on climate change. Perhaps I should not be surprised. While having great respect for McCain, I have observed that when called to do so, politicians, left or right, are perfectly willingly to stick their heads in the sand and ignore reality.

Through the ACIA, growing media attention, and various forms of outreach by the science community, the public was starting to catch on to what was happening in the Arctic. Attention increased after a new record-low ice extent was set in September 2005. The science community was also taking advantage of the growing power of the Internet; one example was the birth in the summer of 2006 of what was to quickly evolve into NSIDC's highly popular *Arctic Sea Ice News and Analysis* site, mentioned back in chapter 2, featuring daily updates of sea-ice extent and regular discussion of evolving conditions.

SHOCK WAVES

Shock waves started building as evidence mounted that the sea-ice cover, already clearly in decline, was now in deep trouble. In August 2005, I participated in a second ARCSS-sponsored retreat, this time at Lake Tahoe, closely focused on where the Arctic was headed in the 21st century. In a paper stemming from the workshop, led by Jonathan Overpeck, we concluded, "The processes and interactions among primary components of the Arctic system, as presently understood, cannot reverse the observed trends toward significant reductions in ice." Stated more simply, a seasonally ice-free Arctic Ocean looked to be inevitable.[16] We were committed.

Concerns grew following new insights into processes driving the sea-ice decline. When the NAO and AO regressed from their high positive states, the enhanced inflow of warm Atlantic-layer waters into the Arctic Ocean had apparently not ceased. But whether this added heat could substantially affect the ice cover had remained in doubt; it was widely believed that the cold halocline would make it too hard to mix this heat upward. Mike Steele had documented retreat of the cold halocline back in 1998, but the halocline subsequently seemed to have recovered. Nevertheless, in 2005 an international team led by Igor Polyakov of UAF presented strong evidence that some of this enhanced Atlantic inflow, which had been occurring in a series of warm pulses, was indeed mixing upward, contributing to summer ice melt and reducing winter ice growth.[17] Shortly thereafter, Koji Shimada of JAMSTEC found that through a complex interaction with declining sea ice, warm water entering the Arctic Ocean in summer from the Pacific Ocean through the Bering Strait was being shunted from the Alaskan coast into the Arctic Ocean, where it fostered further ice loss.[18] So it seemed that, along with a warming atmosphere, the ice was also getting hit by a warming ocean, and from both the Atlantic and the Pacific. It looked to be a triple whammy.

What would the trajectory to the seasonally ice-free Arctic look like? Would it come in fits and starts, or might there be some sort of threshold behavior, or a tipping point, which once passed would result in a rapid, catastrophic loss of the summer ice cover? Marika Holland of NCAR, now a leading voice in the climate-modeling community, provided insight into this question using something known as an ensemble analysis—a series of simulations with the same climate model, but with each one starting from slightly different initial conditions.[19] Some of the simulations showed that as the climate warms and the sea ice thins, a strong kick from natural variability can be enough to propel the albedo feedback process into high gear, such that the overall downward path to a seasonally ice-free Arctic Ocean is interrupted by abrupt plunges in extent lasting a decade or more. This was a disturbing finding. By most accounts, the Arctic system was already rapidly changing, but Marika's results raised the possibility that we could be headed for a cliff.

Shortly thereafter, in May 2007, Julienne Stroeve looked at the trajectory toward an ice-free ocean by comparing the observed rate of decline in September ice extent against hindcasts from a suite of the different climate models participating in the IPCC 4th Assessment (the IPCC report would come out later that year).[20] A hindcast looks at how well climate models can reproduce the observed time series of some variable—in this case, September Arctic sea-ice extent over the period from 1953 through 2006 as well as for the shorter period 1979 through 2006 (corresponding to the modern satellite era). In the real world, the sea ice will be responding to the climate forcings that have actually occurred (e.g., from rising greenhouse gas levels, injection of aerosols into the atmosphere from volcanic eruptions, solar variability). The model simulations include the same forcings as best they can.

The result of the hindcast analysis was that essentially all of the models simulated a loss of September ice extent over the period of observations (implying that yes, the observed loss was at least in part due to rising greenhouse gas levels), but depending on the time window for the analysis, none or very few individual hindcasts showed trends as large as observed (fig. 23). In other words, the models were too slow; compared to what the models were saying ought to be the case, the real Arctic was on the fast track of change. In both the real and model worlds, there is natural climate variability, such as that associated with the AO. Maybe the effects of natural variability on the observed sea-ice record were really, really strong and that accounted for the apparent slowness of the models, but it didn't look that way. Stroeve and colleagues then looked at the forecasts through the 21st century, using a "business as usual" scenario of greenhouse gas growth, at the time probably the most likely of the scenarios that the IPCC was using. Based on the different models, an essentially ice-free Arctic Ocean in September (nowadays generally defined as less than 1 million square kilometers of ice) could be expected anywhere from the year 2050 to well beyond the year 2100. But if the models, as a group, were too slow, this implied that the ice would be gone much earlier.

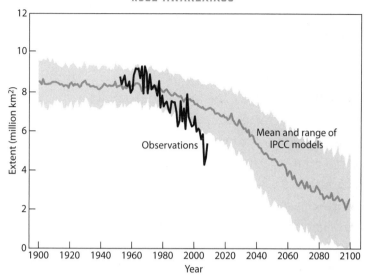

FIGURE 23. The black line and the light gray line show, respectively, the observed September sea ice extent, 1953–2008, and the average September sea ice extent from all of the climate models evaluated by Stroeve's team. The light gray shading shows the standard deviation (spread) between the individual models. This figure is updated a bit and simplified compared to the one originally published by Stroeve and colleagues. Courtesy National Snow and Ice Data Center (NSIDC).

Stroeve concluded, "The Arctic has often been viewed as a region where the effects of GHG [greenhouse gas] loading will be manifested early on, especially through loss of sea ice. The sensitivity of this region may well be greater than the models suggest."[21]

The Summary for Policymakers in the IPCC 4th Assessment that soon followed stated: "Warming of the climate system is unequivocal, as is now evident from observations of increases in global average air and ocean temperatures, widespread melting of snow and ice, and rising global average sea level."[22] It went on to say:

"Most of the observed increase in global average temperatures since the mid-20th century is very likely due to the observed increase in anthropogenic greenhouse gas concentrations. This is an advance since the TAR's [Third Assessment Report's] conclusion that 'most of the observed warming over the last 50 years is likely to have been due to the increase in greenhouse gas concentrations.' Discernible human influences now extend to other aspects of climate, including ocean warming, continental-average temperatures, temperature extremes, and wind patterns."

So, to summarize, the observations were showing that the sea ice was getting hurt by a warming atmosphere as well as a warming ocean, with the hit from the ocean coming from both the Atlantic and the Pacific. Climate-model simulations suggested that, with continued thinning, big plunges in ice extent could be in store. And the observed loss of September sea ice was faster than what the models were saying we ought to have any business seeing. Finally, with respect to the globe as a whole, the long-suspected culprit behind the warming—us— had been fully unveiled. A sense of nervousness started creeping in to the tone of discussions between scientists.

The calendar then turned to 2007. After much planning, the Fourth International Polar Year, or IPY, was launched in March.[23] In the United States, the IPY roughly coincided with the birth of the U.S. Arctic Observing Network, or AON,[24] which, in turn, was closely tied in to SEARCH.[25] AON funded a number of large field projects focused on collecting environmental data to monitor and better understand what was going on in the Arctic. Jamie Morison's North Pole Environmental Observatory, for example, became an AON project. While these events were certainly keys in coordinating Arctic research in the United States and abroad, it is for a very different reason that the year 2007 still stands so tall in the history of Arctic climate research.

The story starts in July. The month began with sea-ice extent below average, but by now, nobody was terribly surprised by that. Then the rate of summer melt quickly picked up steam, and the sea-ice extent started hitting record low values daily. It didn't stop. Through August and September, attention increasingly turned to the North.

By the end of the melt season, Arctic research would never be the same again.

To the utter astonishment of the science community, the sea-ice cover shrank to reach a new record low over the entire period of satellite observation. It didn't just beat the old record set two years earlier in 2005; it completely blew it away (fig. 24). Huge chunks of ice were missing. Late in the summer melt season, the fabled Northwest Passage opened up. There are actually a number of passages through the channels of the Canadian Arctic Archipelago. Of special note was that the deep-water, northern passage, entered from the west through the M'Clure Strait, appeared to open, which would be the preferred route for deep-draft ships. It was the first time in recorded

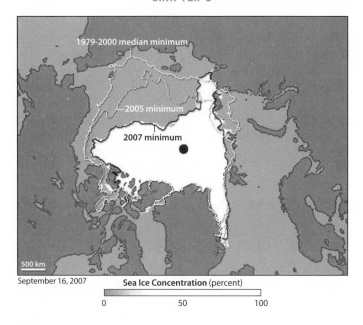

September 16, 2007

Sea Ice Concentration (percent)

0   50   100

FIGURE 24: Arctic sea-ice extent for September 16, 2007, at the September minimum for 2005 (line labeled '2005 minimum') and the median minimum extent calculated over the period 1979–2000. Source: NASA, http://earthobservatory.nasa.gov/Newsroom/NewImages/images .php3?img_id=17800 en:NASA Earth Observatory

history that this had happened. Oddly, at the same time, on the Russian side of the Arctic, the Northern Sea Route was choked by ice north of the Taymyr Peninsula.

A key driver of this record ice loss was an unusual pattern of atmospheric circulation (fig. 25). A persistent area of high pressure at the surface (an anticyclone) parked itself over the Beaufort Sea, north of Alaska. In addition, a persistent area of low pressure parked itself over northeastern Eurasia. The contours on a topographic map connect points with equal elevation. A sea-level pressure map can be read in the same sort of

A Mean SLP June–August 2007 (hPa)

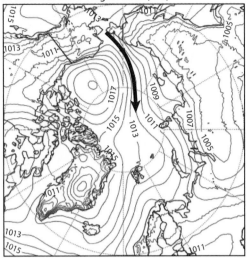

B 925 hPa temperature anomaly June–August 2007 (K)

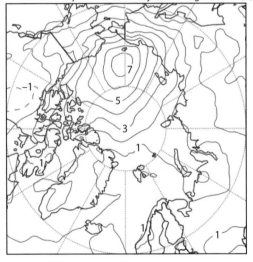

FIGURE 25: Sea-level pressure pattern and anomalies in air temperature at the 925 hPa level (a hectopascal, or hPa, is a unit of pressure) for the summer (June through August) of 2007. The arrow indicates the approximate direction of wind flow at the surface. Property of the author, created by Alexander Crawford at the National Snow and Ice Data Center.

way, except that the contours connect points of equal pressure. Building on our earlier discussion, winds around an anticyclone (the high-pressure zone) blow approximately (not entirely) parallel to the contours of sea-level pressure in a clockwise fashion (in the Northern Hemisphere). Winds around a low-pressure region blow approximately parallel to the contours of sea-level pressure in a counterclockwise fashion. The strength at which the winds blow is, in turn, proportional to the spacing of the pressure contours—the closer the spacing, the stronger the winds. The pattern that developed in the summer of 2007 led to a tight pressure gradient between the high- and low-pressure centers, meaning strong winds, with a component from the south. Winds from the south are warm winds.

This turned out to be a perfect setup to get rid of ice, and later analysis dubbed it the exemplar of the "Arctic Dipole Anomaly."[26] The strong, warm winds from the south brought a great deal of heat into the region between the high- and low-pressure systems, as seen in the pattern of summer air temperature anomalies—that is, departures from average conditions (bottom panel of fig. 25). Hence, a lot of ice melted. The relationship between the area of ice loss relative to the 1979–2000 ice edge (fig. 24) and the area of warm, southerly winds is obvious. The ice was simply eaten away. Because the winds were from the south, ice was also transported away from the shores of eastern Siberia and Alaska poleward. Also, on the Atlantic side of the Arctic, the pattern of pressure and winds helped to transport sea ice out of the Arctic through the Fram Strait into the Atlantic, where it subsequently melted.

It was a crazy year in other ways as well. As sea-ice extent shrank to a new record low, there was an extreme surface melt event over the Greenland ice sheet, 60% greater than the previous high mark recorded in 1998.[27] Then there was a record-high discharge from the combined six largest Eurasian Arctic rivers, accentuating the long-term trend and further arguing for an intensification of the Arctic's hydrologic cycle.[28]

For many scientists, what happened in 2007 ripped away any remaining shred of doubt that a new Arctic was upon us. Sea-ice scientist Julienne Stroeve of NSIDC put it this way: "Now, in our paper published in 2007 [using data from 1953 through 2006], we demonstrated that while essentially all of the climate models were showing that we ought to be losing sea ice, as a group the September ice loss from the models was too slow compared with what we were actually observing. Could part of that be because natural variability has been an especially big player in the observed ice loss? Maybe so, but the effects of natural variability are also captured in the models. Then along comes 2007. Sure, the ice loss that we saw was clearly related to an unusual atmospheric pattern, but the ice cover just wasn't supposed to respond like this."[29]

Sea-ice scientist Dirk Notz, then a postdoc at the Max Planck Institute for Meteorology in Hamburg,

Germany, had a very personal take: "In July 2007, I was co-hosting the IPY [International Polar Year] sea-ice summer school at the university in Svalbard, Norway, at 78°N. The record ice loss of that summer was already apparent at the time of the school, and the urgency of learning more about what was going on was compelling. This got even more apparent as I was afterward joining a two-months-long expedition on a small historical sailboat, which was taking me and others north to allow for the sampling of the retreating ice pack. It was stunning to see that we had to sail all the way up to 82°N to find any sea ice—and it was even more stunning to afterward find that we had apparently been the only ship in the entire area taking measurements that year. As I returned, I couldn't tell which was stronger: the scientific curiosity to understand what was going on, or the sadness of seeing this landscape disappear in front of our eyes."[30]

On a personal note, what I remember the most was a sense of morbid fascination. By this time, I viewed myself as pretty much an old salt on the topic of the Arctic's shrinking sea-ice cover, but I'd never seen anything like this before. While Dirk and a lucky few others had the unique opportunity to be up in the Arctic to see what was happening firsthand, just sitting at our desks we could track what was going on from online images created from the satellite passive microwave data. The University of Bremen had previously set up a great Web site where they posted detailed daily images from the AMSR-E sensor in false color, enabling a blow-by-blow account of the carnage. Much like old baseball fans are in the habit of turning the morning newspaper to the sports section to look at the league standings and box scores, my daily routine became one of getting some coffee, looking at the daily updated graph of ice extent at the NSIDC site, then logging onto the Bremen site to get a sense of the details.

Everyone understood that a big factor was the weather pattern, but we were astounded by what they saw. The postmortem began almost immediately, and blame started to fall on the role of sea-ice thickness. It was known that the ice was getting thinner, and since it simply takes less energy to melt out thin ice than it does thick ice, the ice will respond more strongly to unusual weather events like those in 2007 than it used to. If the same sort of weather pattern had set up 30 years ago, with the ice being so much thicker, it would have been able to take the punch. September extent would certainly have been low, but nothing like 2007. Hence, 2007 provided a great example of how natural variability (the persistent summer weather pattern) and climate change get conflated—as the climate changes, so does the sensitivity of the ice to weather events. In years to follow, work by Rebecca Woodgate at the University of Washington showed that the record low sea ice might also have been linked to an unusually big pulse of ocean heat coming into the Arctic from the Pacific through the Bering Strait, perhaps related to the strong winds from the south in the Bering Strait region.[31]

Would 2007 be the start of a rapid plunge in ice extent like the type that Marika Holland's climate model experiments suggested? That year, I went on record saying that it was very reasonable to expect a seasonally ice-free Arctic ocean by the year 2030. Ice extent for September 2008 was higher than for 2007, albeit still the second lowest in the satellite record up to that time. At a press conference, I was quoted as saying that the Arctic sea-ice cover was in a "death spiral" and that the sea ice had probably indeed reached a tipping point. A story was published in June 2008 in the British paper *The Independent* with the headline "No Ice at the North Pole," featuring a number of quotes from me. The idea of no ice at summer's end at the geographic pole had some basis given how far north ice-free conditions had extended in September 2007. Additionally, Canadian scientist Dave Barber reported that the Arctic Ocean was covered by a lot of thin and rotten ice, as seen by direct observations during a 2008 cruise of the Canadian icebreaker *Amundsen*. Unfortunately (and perhaps intentionally), some interpreted the possibility of no ice at the geographic North Pole to mean that the Arctic Ocean as a whole would have no ice at all. My statements got a lot of attention (especially the death spiral thing), particularly from the climate change skeptics, which led to a fair bit of unflattering e-mail. Some scientists also thought that my statements were over the top.

It seems that I had changed from a fence-sitter regarding human-induced climate change to a loose cannon in a span of five years. Not that I intended to be a loose cannon. An important lesson that I learned from the death spiral interview is to be very careful when speaking to the media—things you say in passing that to you may seem peripheral to the real story may end up being the ones that get the attention and traction. Having said this, given how events have subsequently unfolded, the death spiral analogy seems to have been apt, and others have picked up on it.

Ice extent was higher still in 2009, and some breathed a sigh of relief. But then it dropped again in 2010, and 2011 was even lower (claiming the new number-two spot). While 2007 was apparently not a tipping point, and the death spiral statement was perhaps premature, it was certainly hard to argue that the sea ice was recovering. Indeed, at the close of 2011, the five lowest September extents had all occurred in the previous five years. While some of these years, notably 2008, had a summer atmospheric circulation with some elements of the Arctic Dipole Anomaly, none were associated with the extreme and persistent weather pattern seen in 2007. So the idea of a rapid plunge was still an open issue.

CRANKING UP THE THERMOSTAT

Recall from chapter 2 that there are a number of different processes driving Arctic amplification, and one of the important ones is sea-ice loss. To briefly reiterate but also to expand a bit, as the climate warms, areas of dark, low-albedo open water form earlier in spring and summer, and these dark areas absorb a great deal of the sun's energy, but without a big temperature change. When the sun sets in autumn, much of the heat that had been gained in the upper ocean over the ice-free areas goes back to the atmosphere via sensible and latent heat fluxes (forms of vertical turbulent energy transfer; a bigger latent heat flux puts more water vapor into the atmosphere) and longwave radiation, keeping it fairly warm. Because the atmosphere has a higher temperature, and also has more water vapor (a greenhouse gas), there is also a bigger warming longwave radiation flux back to the surface. Figure 26 shows a schematic, contrasting the low sun period (autumn and winter) under an unperturbed climate versus one with a positive (warming) climate forcing.

In 2006, after looking at all of the observational evidence, including the earlier work by Igor Polyakov, Jim Overland, and Ola Johannessen, and also looking at projections of Arctic temperature change from climate models for the near future (2010–2019), Jen Francis and I concluded that the long-awaited Arctic amplification signal over the ocean due to sea-ice loss was going to

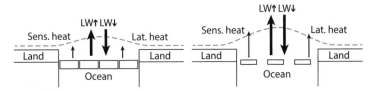

FIGURE 26: The surface energy budget of the Arctic Ocean, contrasting the situation during the low-sun period (autumn and winter) for the Arctic before any human influence on climate (left) and in response to a positive (warming) climate forcing that results in decreased ice extent and thickness (right). The extra heat gained by the upper ocean in spring and summer is released upward in autumn and winter; as shown by the dotted line, this bows isotherms (lines of equal temperature) upward—that is, a dome of fairly high temperatures develops over the open-water areas. The extra heat that is released upward is from stronger transfers of longwave radiation (LW), sensible heat (sens. heat), and latent heat (lat. heat). From Serreze, M.C., and R.G. Barry (2011), "Processes and Impacts of Arctic Amplification: A Research Synthesis," *Global and Planetary Change*, 77, 85-96.

shortly emerge. All that was needed was to get rid of some more ice.[32]

With only a few more years of data, it became clear that Arctic amplification had indeed emerged. I led another little study in 2009 looking at Arctic amplification using data from two atmospheric reanalyses focusing on the period 1979–2007, which included the record-low sea-ice year.[33] First, we showed that the NCAR-coupled climate model (known as CSM) that Marika Holland had been working with projected the biggest Arctic amplification over the Arctic Ocean, most prominent not during summer, but during autumn and early winter. The reanalysis data revealed a strong autumn warming over the Arctic Ocean, biggest at the surface, focused over areas with the biggest sea-ice losses. It was pretty much spot-on with what CSM and other climate models were projecting. We saw it as a smoking gun.

Only a little later, James Screen, an up-and-coming climate scientist with a great deal of energy, then at the School of Earth Sciences at the University of Melbourne in Australia, along with Ian Simmonds, took a closer look using data from the ERA-Interim reanalysis and station data poleward of 70°N.[34] ERA-Interim, produced by the European Centre for Medium-Range Weather Forecasts, was one of a new generation of atmospheric reanalyses that were then coming out; improved in comparison to earlier efforts, it was termed "interim" because it was viewed as a stopgap, to be replaced by an even more advanced product. James Screen's analysis using ERA-Interim showed striking seasonality in the temperature trends computed over the period 1989–2009 (fig. 27). As expected, the trends are smallest in summer and biggest in autumn and early winter, and very importantly, there is a characteristic time lag between the timing of the sea-ice minimum (September) and the biggest temperature trends (October through January) reflecting the big heat loss from the ocean once the sun goes down. While there is a lot more to Arctic amplification than sea-ice loss, these results were still exciting because they provided convincing evidence that the projections from the climate models were being realized. The possibility that the models might be too slow regarding the rate of sea-ice loss didn't change this conclusion.

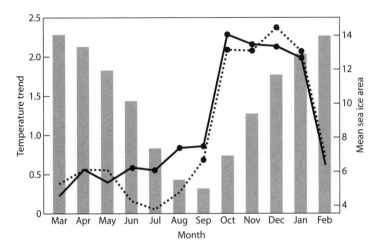

FIGURE 27: The annual cycle of surface temperature trends, 1989–2009, compared to average sea-ice extent. Trends, in degrees Centigrade per decade, are shown by month averaged from meteorological stations north of 70°N (solid line) and from ERA-Interim averaged north of 70°N (dotted line). Solid circles show trends that are statistically significant. The gray bars show the annual cycle of mean sea-ice area (106 sq km). Source: Screen, J.A., and I. Simmonds (2010), "Increasing Fall-Winter Energy Loss from the Arctic Ocean and Its Role in Arctic Temperature Amplification," *Geophysical Research Letters*, 37, L16707. doi: 10.1029/2010GL044136.

The paleoclimate community had also been hard at work trying to place recent Arctic warming into the context of the past. Among other notable efforts, Darrell Kaufman of Northern Arizona University and an international group (that included Ray Bradley) came through with a reconstruction of Arctic summer temperatures over the past 2000 years.[35] Among other data sources, they used paleoclimate data from Arctic lakes, like the lakes that Mike Retelle and Ray Bradley had recovered cores from way back in 1982 and 1983 on Ellesmere Island. Following the now widely recognized "hockey stick"

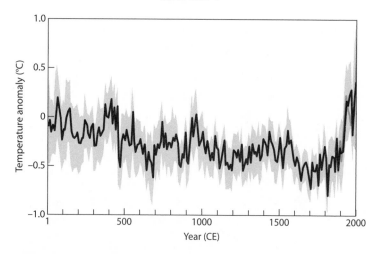

FIGURE 28: The solid line shows a reconstruction of summer Arctic land temperatures over the last 2000 years, based on a composite of proxy records from lake sediments, ice cores, and tree rings relative to a 1961–1990 reference period (gray shading represents variability among different Arctic sites). Adapted from Kaufman, D.S., D.P. Schneider, N.P. McKay, et al. (2009), "Recent Warming Reverses Long-Term Arctic Cooling." *Science,* 325, 1236-1239. Reprinted by permission of AAAS.

pattern, their analysis showed that for the first 1900 years of the record, summer Arctic temperatures were variable, but there was overall cooling, reflecting the Milankovitch forcings discussed in chapter 1. This cooling then ended abruptly with sharp warming in the 20th century, with four of the five warmest decades occurring between 1950 and 2000 (fig. 28).

As already mentioned, in the summer of 2007, as sea-ice extent shrank to a new record low, there was an extreme surface melt over the Greenland ice sheet. In 2009, a team led by Michiel Van den Broeke of Utrecht University, in the Netherlands, attributed about half of the recent mass loss from Greenland to surface runoff (directly from melt) and half to ice sheet dynamics.[36] The accelerating flow of some of the big glaciers draining the ice sheet, including Jakobshavn, Kangerdlugssuaq, and Petermann, drew continued attention. In 2012, Andy Shepherd, of University College London, coordinated input from a huge team of scientists from around the world to assemble all of the information available to date. His goal was to put together a time history of the mass balance of the Greenland and Antarctic ice sheets over the period 1992–2011.[37] Over that time, they found that both ice sheets had been losing mass, hence contributing to rising sea level, with the largest contribution from Greenland. Notably, it became evident that the mass loss from Greenland had also accelerated (fig. 13).

Shepherd's study, which came out in November 2012, ended up being a capstone on what turned out to be yet another a remarkable year.

In the summer of 2012, nearly the entire Greenland ice sheet saw surface melt. Remember that the percent of the ice sheet experiencing surface melt had shown a general increase through the years, consistent with a warming climate. Now the melt extent area had essentially hit the limit of the chart and could go no higher. As the summer wore on, it became obvious that another new record low in sea ice was in store, and when the final numbers came in that September, the old 2007 record that had astounded the science community hadn't just been beaten, it had been shattered (fig. 6). Remarkably, the new record had been set without a very highly favorable summer atmospheric circulation pattern like that seen in 2007. Yes, it was a warm summer and the atmospheric pattern was by no means unfavorable, with elements of the Arctic Dipole Anomaly, but apparently, by 2012 the ice had gotten so thin in response to both a warming atmosphere and a warming ocean that we got a new record without a perfect storm.

A NEW ERA

By the end of 2012, the major pieces of the puzzle of the changing Arctic had pretty much fallen into place. It had taken at least two decades to fully appreciate how strongly the Arctic's atmosphere, land, and ocean are influenced (and will continue to be influenced), by aspects of natural variability. With the AO and NAO playing such a strong role in this variability, the human fingerprint of change that so many of us were looking for was in many ways largely hidden. Indeed, it is now fairly clear that the enhanced influence of Atlantic derived waters—the very thing that attracted so much attention to the Arctic in the early 1990s—has very strong expressions of decadal-scale variability. While at least, in part, this is itself related to the behavior of the AO and NAO, there is more going on that remains to be understood. But with the passage of time, the human fingerprint in the Arctic, as expressed in terms of its sea-ice cover, temperature, permafrost thaw, and other variables, finally emerged above all of this noise, and once it did, it reared up and roared.

We have also learned that natural climate variability and greenhouse warming can themselves be entangled; for example, because the sea-ice cover is now thinner than it used to be, it now responds differently to natural variability of both the atmosphere and the ocean. We've learned that as it warms up and surface melt over the Greenland ice sheet increases, the flow of the massive glaciers that drain the ice sheet accelerates. The long-awaited Arctic amplification is here in a big way, but it seems to be a much more complex animal then we ever could have imagined.

7

LOOKING AHEAD

It is a foregone conclusion that in future generations, whether on the ocean or on the land, the Arctic will have much less ice. Winter darkness will still bring low temperatures, and with them snow and ice, but this winter cold will have significantly faded. Perhaps with the exception of the higher-elevation areas of the Greenland ice sheet, the snow that falls, and the ice that grows in winter, will not survive the stronger summer warmth. That the Arctic Ocean will become free of sea ice in late summer and early autumn is a given; the only question is how quickly it will happen, which will depend on the relative roles of a warming atmosphere and a warming ocean, the vagaries of natural climate variability, and how quickly greenhouse gas concentrations continue to rise. As this book goes to press, some of the biggest remaining unknowns revolve around the impacts of a transformed Arctic. When will the permafrost carbon feedback kick in, and will it be strong enough to push global temperatures even higher? Will a much warmer Arctic have significant impacts on weather patterns in lower latitudes? Will this affect agricultural patterns? How quickly will sea-level rise due to melt of the Greenland ice sheet, Arctic ice caps, and glaciers? Given prospects of increased shipping and extraction of resources, how much busier will the Arctic become, and will this lead to conflicts? These are questions that should concern us all.

SAFE BETS

The IPCC 5th Assessment was finalized in 2013, and it left no doubt that humans were hitting the climate system with a ball-peen hammer. As stated in the Working Group I Summary for Policymakers: "Human influence on the climate system is clear. This is evident from the increasing greenhouse gas concentrations in the atmosphere, positive radiative forcing, observed warming, and understanding of the climate system." Furthermore, "Human influence has been detected in warming of the atmosphere and the ocean, in changes in the global water cycle, in reductions in snow and ice, in global mean sea-level rise, and in changes in some climate extremes. This evidence for human influence has grown since AR4 [the 4th assessment]. It is extremely likely that human influence has been the dominant cause of the observed warming since the mid-20th century." Specifically, with respect to the cryosphere: "It is very likely that the Arctic sea-ice cover will continue to shrink and thin and that

Northern Hemisphere spring snow cover will decrease during the 21st century as global mean surface temperature rises. Global glacier volume will further decrease."

Yogi Berra once quipped, "It's tough to make predictions, especially about the future."[1] True enough, but with respect to climate, we can make scientific predictions with ever-increasing confidence. Based on everything that has been learned over the past 30 years, much about the future of the Arctic is a pretty safe bet.

By 2050, safely assuming no drastic reductions in greenhouse gas emissions in the immediate future, the Arctic Ocean will likely have little or no sea ice at summer's end. Given the large natural variability in the Arctic, such as that associated with the NAO, AO, and the Arctic Dipole Anomaly, the first occurrence of an essentially ice-free summer (less than 1 million square kilometers of ice) might have taken place as early 2030 or 2040. Whatever summer ice remains will largely hang out north of the Canadian Arctic Archipelago. This is because the Beaufort Gyre (fig. 14) tends to push ice up against the coast, where it can thicken as ice floes pile up against each other. There will still be winter ice, but it will be thinner than it used to be. It will, of course, be a warmer Arctic, not just because of the direct effects of higher greenhouse gas levels but because of the processes and feedbacks that lead to Arctic amplification. With more water vapor in the atmosphere, precipitation will be higher than today. As has already been observed over many parts of the world, there will be more extreme precipitation events, at least in some areas, with the potential for flooding and erosion. The snow season will be shorter, with snow forming later in autumn and melting out earlier in spring. Vegetation will have changed, with large areas of treeless, windswept tundra taken over by shrubs. Over large areas, permafrost near the surface will have thawed; in areas where permafrost is presently isolated (e.g., confined to north-facing slopes), it will have largely disappeared.

Still, there is a lot out there that isn't completely understood, and surprises may be in store. There are, to borrow from Donald Rumsfeld, "known unknowns," or things we know that we don't really know.[2]

KNOWN UNKNOWNS

How much busier will the Arctic be in 2050? For a long time, there will still be winter ice, so the Arctic Ocean will still provide only seasonal shipping routes. Nevertheless, as sea ice disappears, the Arctic will become more accessible. At that point, whether year-round trans-Arctic shipping makes sense will become mostly a matter of economics, such as the price of oil and natural gas, and the relative costs of shipping through the Suez or Panama Canals.

Royal Dutch Shell was looking for oil and gas in the Chukchi Sea, then didn't find as much as they had hoped for and left. But if prices skyrocket, they may be back.

Maybe lots of tourists will want to pay to take cruises through the Northwest Passage and the Northern Sea Route. Indeed, this is already happening; in the summers of 2016 and 2017, the massive cruise liner *Crystal Serenity*, featuring 13 decks, eight restaurants, and a spa, cruised through the Northwest Passage without a hitch. A busier Arctic implies the need for investment in year-round port facilities and disaster response. Nobody has quite figured out how to deal with massive oil spills in Arctic waters, or the logistics of evacuating 1000 tourists from a stricken luxury cruise ship in the remote waters of the Northwest Passage. How busy the Arctic gets also depends on geopolitical events. Perhaps terrorists will disable the Suez or Panama Canals. Perhaps Vladimir Putin or his successor will claim the Arctic Ocean as part of his or her territory and put up a tollbooth in Murmansk.

There still will be plenty of winter ice for a long time, and ice conditions even in summer will remain variable. Hence, there is an increasing, not a decreasing, need for capable icebreakers. Russia has invested heavily in the Arctic, and in 2016 it launched a new super-duper nuclear-powered icebreaker the *Arktika*. The United States has woken up to the need for a new icebreaker fleet to be operated by the U.S. Coast Guard; discussion is ongoing over the mix of vessels that should be built. All the United States really has right now is the *Healy*, commissioned in 2000, and the aging *Polar Star*, commissioned way back in 1967.

A further challenge to marine operations is that only perhaps 10% of the Arctic Ocean seafloor has been charted to modern standards.

The fishing industry will likely also change. In 2016, delegates from nine countries and the European Union met in Washington, D.C., to discuss a U.S. proposal to put the central Arctic Ocean off limits to fishing until enough research has been done to inform sensible regulations.[3] As noted back in chapter 2, the issue here is that as the ice recedes, commercially important species like cod and haddock will move north, and at some point, the Arctic Ocean may no longer be an effective barrier separating Atlantic and Pacific species. The proposed moratorium represents a rare case of forward thinking; generally, regulations are put in place only after stocks have dwindled or collapsed.

Some known unknowns are more unknown than others. Some animal species are likely to have a hard time, such as polar bears and walruses, although others may prosper. Moose, for example, should find a shrubbier Arctic quite tasty but caribou may not like it. A known concern with caribou and reindeer is that rain-on-snow events, which are likely to increase in the future, make it difficult for them to forage. Because of more abundant zooplankton, bowhead whales may do well. Indeed, we are already seeing some of these changes, but ecological shifts are tough things to pin down. Ecological systems can also be quite fragile, and even small changes could have big impacts.

Losing the summer sea-ice cover is seemingly inevitable, but when will we get there, and what will it look like? Remember that the observed rate of sea-ice loss was faster than expected based on hindcasts from climate models participating in the IPCC 4th Assessment. More recently, Julienne Stroeve and colleagues looked at whether the next generation of models participating in the IPCC 5th assessment did any better at reproducing the observed September trend. As a group, they did, although many still looked slow, and, as with the earlier generation of models, there was a big spread between different models both with respect to simulated trends over the period of observations (the hindcasts) and the projections of when a seasonally ice-free Arctic becomes reality.[4]

Will the ice thin to some tipping point after which it takes a series of big plunges, as some had read into results from Marika Holland's study? Although sea ice recovered for a few years after 2007, the new record-low extent for September 2012 fueled some voices arguing that a near-complete loss of the summer ice cover might occur in just a few years.

As it turns out, in the period between publication of Marika's paper in 2006 and the record-low ice extent of 2012, the tipping point scenario was dealt a blow. One of the arguments is that after a really big summer ice loss, there is a tremendous heat loss to the atmosphere and then to outer space during the following autumn and winter (when it's dark over the Arctic Ocean), so there is a lot of ice growth and the ice tends to recover.

Remember that the part of Arctic amplification due to sea-ice loss is actually a reflection of this ocean heat loss (fig. 26); on its way out to space, some of the heat lost from the ocean warms the atmosphere. It's a counterintuitive thing; while on the face of it, Arctic amplification linked to ice loss ought to get rid of more ice, it's actually an ocean-cooling process. The heat loss argument helps to explain why really big negative anomalies in September ice extent (such as in 2007 and 2012) tend to be followed by a higher September ice extent the next year (such as in 2008 and 2013); it's the natural negative, or stabilizing, feedback in the system. In early 2011, Steffen Tietsche, then at the Max Planck Institute for Meteorology in Hamburg, Germany, presented some strong evidence for this in a modeling study.[5] Still, it's not a slam-dunk argument; if there is more heat in the upper ocean at summer's end, there is also more to give up before any ice can form. Also, the negative feedback doesn't necessarily mean that the downward trend in September extent won't accelerate as the ice thins and things keep warming up. Indeed, the observations show that this acceleration has probably already happened. So while the death spiral argument may well hold up, there probably isn't a vertical cliff. What a relief!

Dirk Notz, of the Max Planck Institute and coauthor of the Tietsche study, was one of a number of scientists concerned about extreme views that had been floating around. "In 2007, we all started discussing why the ice loss during that summer happened so rapidly. Things were quite different in 2012. This was because in the meantime, a number of claims had been made that predicted the near-immediate loss of the remaining sea ice. Hence, in 2012, rather than discussing the rapidity of the loss, I felt that we had to discuss why the ice loss most likely would be slower than the most extreme forecasts, and that even after the 2012 minimum we would most likely recover a bit in the following years [via the autumn and winter heat loss mechanism]. Rather than being able to solely focus on the still amazing rapidity of the loss of sea ice, we now had to defend our credibility as climate scientists by pointing out that the ice loss would probably be slower than the most extreme forecasts. In that respect, I felt our communication was failing. And I felt sad at realizing how focus shifted to the most extreme, highly unlikely forecasts, rather than simply remaining on the fact that this ice loss was real. And it was extreme, without any exaggeration needed."[6]

Still, much remains to be understood about the downward trajectory and the relative roles of top-down forcing on the sea ice (influences of the atmosphere on the top of the ice) and bottom-up forcing (influences of the ocean on the bottom of the ice). In terms of top-down forcing, recent research has shown that years with low September ice extent tend to be preceded by enhanced transport of humid air from lower latitudes into the Arctic during spring. This is because humid air, with its high water vapor content and association with extensive cloudiness, provides an enhanced greenhouse effect promoting early melt.[7] As the climate warms, there will be more water vapor in the atmosphere, and so more transport from lower latitudes. Summer melt ponds atop the ice cover also seem to be important. Melt ponds are relatively dark and hence readily absorb solar energy, which results in more ice melt via the albedo feedback. Melt ponds that form on hummocky multiyear ice tend to be small but deep, while those that develop on flatter first-year ice are shallower but broader. With more of the sea-ice cover in spring represented by first-year ice, summer melt ponds now cover a larger area, which makes the albedo feedback mechanism all the more effective. Yet other studies argue that the influence of variability in summer weather patterns on the observed downward trend may be bigger than we think, although this is complicated by the observation that as the ice gets thinner, its response to weather patterns changes.[8] Regarding bottom-up processes, recall that pulses of warm water from the Atlantic have been tracked entering the Arctic Ocean. According to the most recent work by Igor Polyakov and colleagues, while the inflow of Atlantic waters has actually slowed since peaking in 2008, its influence on the sea-ice cover has only become more prominent, strongly limiting winter sea-ice growth in the Eurasian Basin of the Arctic Ocean.[9] Igor links this to retreat and weakening of the insulating cold halocline, which then allows for more vertical mixing. So yet again, we encounter complicating aspects of our old friend called natural variability; the Atlantic inflow increases, then backs off, the cold halocline weakens, recovers and weakens again. This increased "Atlantification" of the Arctic, which may well be a temporary thing given what we now know, explains why the Barents Sea region has seen so little winter sea ice in recent years, and, in turn, why there is a big positive trend in the winter surface air temperatures in this area (fig. 10).

Turning to the Pacific side of the Arctic, there is a flow of fairly fresh and warm ocean water into the Arctic Ocean through the Bering Strait (fig. 14). Because the waters in the North Pacific are somewhat fresher than in the North Atlantic and hence a bit less dense (hence a given mass occupies a large volume), sea level is actually a little bit higher in the North Pacific, and the water wants to flow downhill from the Pacific and into the Arctic Ocean. Moorings in the shallow Bering Strait provide a record of the Bering Strait inflow. Recent work has shown that variability in the heat inflow strongly affects sea-ice conditions in the Chukchi Sea,[10] and possibly over a much larger region. There has also been an upward trend in the annual heat inflow.[11] As with the Atlantification, whether this trend will persist, stop, or reverse, remains to be seen.

The future contribution of Greenland, Arctic ice caps, and glaciers to sea-level rise is also in question. As introduced back in chapter 5, glaciologists are still trying to pin down the different processes leading to the observed accelerating flow of the massive glaciers that drain the Greenland ice sheet, such as thinning of the calving tongues that reduces backpressure, and basal lubrication via the Zwally Effect, and the relative role of ice dynamics and the surface mass balance (winter accumulation versus summer melt).

Marika Holland, when asked about pressing research issues, replied, "I think that the biggest uncertainty that is relevant for society is what will happen to the Greenland ice sheet. Most global climate models do not even incorporate active ice sheet components. Even in the models that do have an active ice sheet, the treatment is overly simple. The rate at which Greenland melts and slides away seems a huge wild card with enormous implications."[12] This emphasizes the point that despite the many advances that have been made in climate models, they are still incomplete representations of the real world. They are critical tools in climate research, and we have learned a lot from them, but they are still nowhere near as sharp as we would like.

Adding to the complexity, Greenland itself may be affected by the loss of the sea-ice cover. Arctic amplification linked to sea-ice loss can affect air temperatures over the ice sheet, and more open water in autumn and winter may eventually alter patterns of precipitation over the ice sheet. Extensive sea ice in combination with blocks of ice calved from glaciers (termed mélange) exerts a force against the flow of the glaciers draining the ice sheet; remove this mélange and the glaciers may flow faster.

Two known unknowns that relate to the physical environment—the permafrost carbon feedback and the effects of Arctic amplification on mid-latitude weather patterns—merit some detailed visitation. Of course, there may also be unknown unknowns in the Arctic climate system—that is, total surprises. As far as those go, we'll wait and see what happens.

PERMAFROST CARBON FEEDBACK

Plants take up carbon through photosynthetic activity— this is called gross primary productivity, or GPP. Plants also release carbon through autotrophic respiration as they convert simple sugars—the raw products of photosynthesis—into leaves, roots, and wood. Plants are called autotrophs because they make their own food. The difference between GPP and autotrophic respiration is called net primary production, or NPP. But there is also heterotrophic respiration, or HR, which is the release of carbon by animals, microbes, and other organisms that consume organic matter. Readers of this book are heterotrophs; we all consume autotrophs (because vegetables are good for us), and many of us also consume heterotrophs (because animals are tasty and contain concentrated proteins and fats). Depending on whether the respiration process is aerobic (with oxygen, such as in our case) or anaerobic (without oxygen), the heterotrophic respiration releases either carbon dioxide or methane.

Methane is a carbon atom attached to four hydrogen atoms, and is the simplest of the hydrocarbons. Like carbon dioxide, methane is a greenhouse gas, and as a greenhouse gas, it is actually much more powerful, molecule for molecule, than carbon dioxide, but it is also much less abundant in the atmosphere and has a shorter residence time. Katey Walter, an aquatic ecologist and biochemist from the University of Alaska, has posted a number of impressive videos to YouTube illustrating methane release from Arctic lakes. The methane, a product of anaerobic respiration, bubbles up from lake bottoms and collects under the ice in winter. Katey's approach is to poke a hole in the ice and then, standing back for safety, light off the highly flammable methane, to great effect.

In terms of the overall carbon balance, a key issue is the net ecosystem production, or NEP, which is the overall uptake or release of carbon by the biosphere. Stated simply (as the only equation in this book) NEP = NPP minus HR. Does carbon sequestration from NPP exceed the release of carbon by HR? If so, NEP is positive, so there is a net carbon uptake by the biosphere. Or does HR exceed NPP? If so, the biosphere releases carbon.

Arctic and subarctic soils contain a great deal of carbon locked up in permafrost as frozen organic matter. Estimates suggest that the permafrost holds around

800 gigatons of frozen carbon, roughly twice the amount of carbon presently held in the atmosphere. As long as this carbon is frozen, it poses little concern for global climate. However, as the climate warms, more of the Arctic permafrost will warm and thaw, and so too will the frozen organic matter. When it thaws, the microbes in the soil (heterotrophs) will resume eating the organic matter and then release some of this stored carbon back to the atmosphere, raising the atmospheric load of greenhouse gases. If this net carbon release to the atmosphere is significant, the resulting warming due to the higher concentration of greenhouse gases in the atmosphere will exacerbate the carbon loading linked to the burning of fossil fuels, meaning more permafrost thaw, more carbon release to the atmosphere, more warming, and so on. This is the permafrost carbon feedback, or PCF (fig. 29).

Little more than a decade ago, the possibility of a significant PCF wasn't on the radar screen, but the problem has received a great deal of attention since then, mainly through modeling studies. An effort led by Kevin Schaefer, a permafrost scientist at NSIDC, suggests that sometime around the year 2030, depending on the assumptions made, permafrost regions may turn into a net source of carbon to the atmosphere, and that by the year 2200, this could lead to a total equivalent increase in atmospheric carbon dioxide of almost 90 parts per million.[13] As of 2017, the atmospheric carbon dioxide concentration was just about 405 ppm.

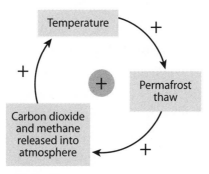

FIGURE 29: The permafrost carbon feed-back. Source: ttp://cars-kill.weebly.com /carbon-dioxide.html

Concerns have also been raised over the potential for massive releases of fossil methane from submarine sediments that underlie the shallow continental shelf seas of the Arctic Ocean. Under the right conditions of both temperature and pressure, methane can become trapped in the sediments in a solid clathrate form—basically, as an ice that burns. The thinking is that as the ocean warms up (and it is), the hydrates will warm and melt, and massive releases of methane (natural gas) may ensue. This is definitely disaster movie material.

There have been observations of plumes of methane gas bubbles rising from the seafloor in the west Spitsbergen continental margin and from the East Siberian Sea. This has been argued to be a response to ocean warming, reducing the extent of the zone at which the clathrates can survive, but it could also be seeps from the natural gas reservoirs that are abundant in these areas. A number of criticisms have been raised. For one, there are no systematic, long-term observations of methane releases from the seafloor, so it's hard to put recent observations into a meaningful context. It has also been argued that any methane release will tend to happen fairly slowly, as opposed to a bomb. However, the idea of a massive release finds support in some modeling studies, and it has been maintained that the release of only a small fraction of methane along the East Siberian Arctic shelf region (the Laptev and East Siberian Seas and the Russian part of the Chukchi Sea) could trigger an abrupt climate warming.[14] Sea-ice loss is viewed as an important player in this. For one, sea ice serves as a barrier that limits methane emissions to the atmosphere; lose the sea ice and the barrier is gone. Also, with less sea ice, there is more wave action, which can initiate mixing, ventilating bubble-transported and dissolved methane from the water column, favoring high rates of emission to the atmosphere.

Natalia Shakhova of IARC has been one of the strongest voices on the topic, and while acknowledging the many uncertainties, she urges the need for continued study: "Subsea permafrost is the most troublesome Arctic component in terms of possible climate change consequences. Subsea permafrost has begun to lose its integrity, and migration pathways for gas preserved within and beneath it have begun to open. A massive methane release from the East Siberian Arctic Shelf, which is estimated to hold more than 80% of existing world subsea permafrost and permafrost-related Arctic shallow shelf hydrates, is the likeliest mechanism to cause dramatic climate change."[15]

ARCTIC AMPLIFICATION AND WEATHER PATTERNS

An issue that has gotten a great deal of attention, not just in the science community but in the media, is the idea that Arctic amplification, by changing the temperature gradient in the atmosphere between the Arctic and lower latitudes, will change the behavior of the jet stream, which, in turn, will be manifested as a change in weather patterns in the middle latitudes. The reason for the media attention should be obvious; weather is something we can all relate to, and blaming a warming Arctic for the blizzard you might be experiencing has a certain attention-grabbing aspect. Unfortunately, the proposed link is also prime fodder for climate change alarmists and skeptics alike.

To fully appreciate the Arctic-mid latitude link requires understanding a little bit more about atmospheric dynamics, which can be done with the aid of figure 30. The sun's rays strike the earth's surface more directly in the lower latitudes than in the higher latitudes. This is why it is warmer in Hawaii than in Barrow, Alaska. This equator to pole (latitudinal) temperature gradient holds not just at the surface, but through much of the depth of the atmosphere. Because there is a temperature gradient in the atmosphere, there is also a pressure gradient in the atmosphere. Warm air is less dense than cold air, so if one measures the air pressure at some fixed height (say, 6 kilometers) above sea level, the air pressure will be higher in the lower latitudes and lower in the higher

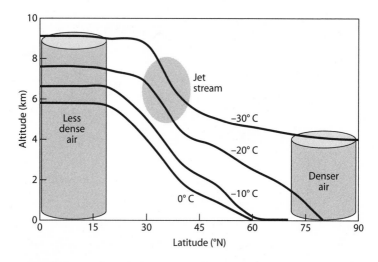

FIGURE 30. Lower latitudes are at the left, higher latitudes to the right; the jet stream, in approximate geostrophic balance, is blowing out of the page toward the viewer. The figure, based on data from an atmospheric reanalysis, blends the polar jet stream with the subtropical jet stream that lies to the south, but gets across the key concept that the jet stream is related to the temperature gradient. Property of the author, created by Alex Crawford, National Snow and Ice Data Center.

latitudes. As a result, the physical thickness (in feet or meters) of a column of air of a given mass (a column containing a given number of gas molecules) stands taller in the lower latitudes. This then means that above the fixed level in the atmosphere under consideration, more of the mass of the atmosphere is above us in the lower latitudes, so at the level being considered, the pressure that the air exerts is higher (the air pressure at a given location is simply the weight, per unit area, of the overlying air).

The pressure gradient will initiate a motion of air from higher to lower pressure (south to north in the

Northern Hemisphere, or left to right in fig. 30)—that is, the pressure gradient initiates a wind. However, because the spherical earth rotates on its axis, there is a Coriolis force that turns the air to the right of its motion in the Northern Hemisphere.[16] The end result is that in the middle latitudes and higher latitudes, the primary motion of air at our fixed level above the surface—that is, the wind—will tend to primarily blow not down the pressure gradient (south to north), but parallel to it (west to east, or out of the page in fig. 30). This situation is called geostrophic balance, which is a balance between the pressure gradient (south to north, acting to the left when your back is to the geostrophic wind) and the Coriolis force (north to south, acting to the right when your back is to the geostrophic wind). The strength of the geostrophic wind depends on the strength of the pressure gradient: the stronger the pressure gradient the stronger the wind. What happens is that a stronger pressure gradient will make the wind blow more strongly from south to north, but this also means a stronger Coriolis force acting to the right, so a geostrophic balance is achieved with a higher wind speed. For reasons we don't have to get into here, the temperature (hence pressure) gradient from south to north is not uniform but is most concentrated over a fairly narrow latitudinal band. Hence, there is a concentrated band of winds high in the atmosphere called the polar jet stream, marking the separation between the colder air to the north and warmer air to the south. Approximate geostrophic balance is the same reason why winds blow clockwise around high-pressure cells (anticyclones) and counterclockwise around low-pressure centers (cyclones) at the surface (fig. 25).

Finally, and this is important, the polar jet stream is not just a straight-line band of strong winds blowing from west to east at high altitude, but rather is wavy. In some areas the jet steam dips to the south, and in others it extends into higher latitudes, forming what are known as troughs and ridges. These patterns of troughs and ridges are known as Rossby waves, after Carl Gustav Rossby, who first successfully described their behavior. There are long waves that span continents, and shorter ones embedded within them. The pattern of Rossby waves is always changing, and storms (cyclones) and high-pressure areas (anticyclones) at the surface form along preferred areas of these Rossby waves. The colder air to the north of the jet stream is termed the polar vortex. Cold snaps such as occur every winter in places like Chicago occur when the jet stream dips south to put the region in the cold air within the polar vortex. Unfortunately, the media has somehow gotten into its mind that the polar vortex is some sort of malicious beast; hence headlines along the theme of "Chicago bracing for invasion of the latest polar vortex."

It hopefully now makes some sense that, if the Arctic strongly warms through removing the sea-ice cover, the south-to-north temperature gradient will change, changing the pressure gradient and, in turn, changing the jet stream. Since the Arctic amplification due to ice loss is biggest in autumn and winter, this is also when the effects on the jet stream, and hence surface weather patterns, ought to be most prominent.

Jen Francis and colleagues noted that patterns of atmospheric circulation following Septembers with low sea-ice extent tend to be different than those following Septembers with high extent, which they interpreted as being partly driven by effects of the contrasting ice conditions on the atmosphere.[17] This provided observational evidence to back up conclusions from previous experiments (and many that followed) with climate models showing that patterns of atmospheric circulation beyond the Arctic might indeed respond to changes in the sea-ice cover. These tended to be idealized sensitivity experiments; artificially remove the entire ice cover, or the ice cover from some part of the Arctic Ocean, and see what happens. It might be thought that if a warmer Arctic reduces the temperature gradient, it would simply lead to a weaker jet stream, but the situation seems to be much more complicated. For example, modeling studies and other observational efforts suggest that where the sea ice is removed is important. Losing ice in the Kara and Barents Seas region, for example, seems to lead to a different atmospheric response than losing it in the Beaufort and Chukchi Seas. Couplings between the stratosphere and the surface also seem to be important.

In 2012, Jen Francis and Steve Vavrus teamed up and published an observationally based study (using data from an atmospheric reanalysis) linking Arctic amplification to extreme weather events in middle latitudes.[18] They argued that Arctic amplification affects the amplitude (elongation) of the Rossby waves (really deep troughs and really strong ridges) and how they move, and that this could result in "stuck" weather patterns. For example, one might get a situation in which the Rossby wave pattern gets stuck so that the eastern half of the United States is in a winter deep freeze with lots of snow ("snowmageddon," associated with an "invasion of the polar vortex"), while at the same time it is absurdly warm over Alaska.

Their study caused quite a stir and attracted its share of snarky remarks from global warming skeptics. Global warming puts Chicago in a deep freeze? What have these climate scientists been smoking? There was also pushback within the science community. Some of this criticism focused on perceived flaws in the data analysis,[19] and some of it questioned the basic premise of a control on the stuck weather patterns by Arctic amplification. For example, while it is readily acknowledged that sea-ice loss leads to strong warming at and near the surface of the Arctic Ocean, it seems that the warming from ice loss doesn't extend through a deep enough layer of the atmosphere to have a really big effect on the atmospheric circulation, and that the strong Arctic warming observed deeper in the atmosphere is actually more of a result of variations in sea surface temperature outside of the Arctic, in turn driving changes in atmospheric circulation that affect Arctic temperature.[20,21] In other words, Francis and Vavrus were putting the cart before the horse. Other recent work argues that the stuck weather patterns just represent natural climate variability. As of the publication of this book, the debate continues, but the weight of evidence increasingly supports a significant Arctic influence.

It is widely acknowledged, however, that as air near the Arctic surface continues to warm, wintertime cold air outbreaks, when Arctic air plunges south to affect middle latitudes (invasions of the polar vortex, as the mainstream media might say), will not be as severe as they used to be. Some of my strongest memories of growing up in the state of Maine in the 1960s and 1970s are from when winter cold fronts would barrel through, heralding the arrival of Arctic air and subzero temperatures that seemed to last for weeks and make the snow squeak. I can't pin down exactly why those memories are so sharp and so fondly held; maybe it just comes from a lifelong fascination with primal forces of nature. It's still hard for many of us to accept that we are now in control of our planet's climate, and that if we do nothing, the joys of winter cold that I remember will slowly fade. But the Arctic tells no lies.

EPILOGUE

The motives of the Arctic explorers of the 18th, 19th, and early 20th centuries included the pursuit of wealth, finding a shortcut to the riches of the Orient, attaining personal fame and glory, and discovering new lands, as well as pure scientific discovery. The explorers all needed maps of the Arctic and an understanding of Arctic climate, and this meant building a knowledge base, with each expedition drawing and building on what had been learned by those who preceded them. It was a messy and uneven process, and filled with tragic events, such as John Franklin's expedition to find the Northwest Passage in the *Erebus* and *Terror*, and Greely's ill-fated march to the south after evacuating his camp at Fort Conger after the First International Polar Year. But even by 1920, it is impressive how much was known about the Arctic.

Today, we have a remarkably mature, albeit still incomplete understanding of how the Arctic works, how it has changed, and where it is headed. There are still many unknowns, but, just as the science community strove to understand the changes that started unfolding in the early 1990s, so will the community seek to find the answers. And this is because scientists like those in this book, whether they are observationalists, modelers, or both, and whether they are climate scientists, oceanographers, glaciologists, or ecologists, are also explorers, just of a different mold than the Arctic explorers of the past. The same drivers are there—to discover and to understand—and like the explorers of old, scientists also stand on the shoulders of their predecessors. Scientists are just asking different questions and using different tools and methods.

New scientific observations and analyses regularly expose weaknesses and flaws in existing ideas, which are then either modified or replaced by new ones. But as in any human endeavor, the scientific process is prone to human frailties, including vanity, envy, competition, greed, and narcissism. Anyone who claims that these things don't things exist in science is either lying or willfully ignorant. All scientists have colleagues whom they work well with, and there are those they avoid. Shouting matches at conferences between scientists with different views are not unheard of. The peer-review process can get testy at times and remains imperfect. Papers making strong contributions sometimes get rejected because of a seemingly unwarranted bad review. Some papers that seem to offer only minor insights fly through review and are published in prominent journals. It can be a mystery. And as we have seen, science is not divorced from politics. But science endures. It is a competitive, regenerative process in which only the best ideas, concepts, and theories survive the test of time.

Pure scientific research cannot be done in the absence of public support; the taxpayers like you, who fund organizations like NSF, NASA, and NOAA, and the work of those who oversee the funding process. The future of science funding is a cause for concern. Due to budgetary pressures, funding for Arctic research has in general been rather flat in recent years. Numerous AON projects have ended. As one prominent example, in 2015, after a long run that started in 2000, the North Pole Environmental Observatory was shut down. As Jamie Morison laments, "With the canceling of the project, we are losing some key repeat observations that will be impossible to duplicate in the future." However, other AON projects have survived, and others will come online. A big new international project with a lot of European backing called the Multidisciplinary Drifting Observatory for the Study of Arctic Climate [MOSAiC]—essentially SHEBA on steroids—will soon be getting underway. The U.S. science community also has some cause to be optimistic, for in 2016, NSF director France A. Córdova announced Navigating the New Arctic, an initiative that will include an expanded observation network, as part of a broad plan to shape the agency's future. Then again, as this book was being written, an administration that was less receptive to the science of climate change stepped into the White House.

Some of the scientists who spent their careers trying to understand the evolving Arctic have died; some have retired, while others have moved on to different things.

Many are still very active, and just as curious as they were back in the early 1990s when the first hints of Arctic change began to emerge. Younger generations of scientists, armed with new and powerful tools, will continue in the quest to understand.

And there is plenty of work to do, for the Arctic is changing at an accelerating rate, punctuated by jaw-dropping events. The winter of 2015/2016 saw what seemed to be an unprecedented heat over the Arctic Ocean. At the very end of December 2015, there was a brief period when the surface air temperature at the North Pole appears to have actually risen to above freezing. Yes, Arctic amplification has been part of the new Arctic for at least a decade now, but having late-December temperatures at the North Pole at or above 0 degrees Celsius is simply unheard of. The heat wave persisted, slowing the wintertime growth of sea ice, and on March 24, 2016, when Arctic sea ice reached its seasonal maximum extent, it was the lowest maximum ever recorded, besting the previous record set just a year earlier in 2015. Then there was another heat wave in the autumn and early winter of 2016, in some ways even more impressive than the one seen a year earlier, and both October and November saw record lows in ice extent. There was even a brief period in the middle of November 2016—normally the period of most rapid seasonal ice growth in the Arctic—when ice extent actually declined.

It's never wise to read too much into individual extreme events, but what has recently been observed has gone beyond reasonable boundaries. It qualifies as crazy. As Seth Borenstein, a longtime science journalist with the Associated Press, lamented to me in a telephone conversation in early January 2017, he's running out of stories about the Arctic—how many times can a journalist report on what is happening in the Arctic before it becomes so repetitive that people lose interest?

My own personal journey is far from over, and it increasingly seems that the mission I have taken on is to communicate the science and make sure that society does not lose sight of the importance of what is happening in the Arctic. If this book has helped to open some eyes, then I consider it to be a success. Like many scientists, I'll probably never fully retire—science gets in your blood. It's what gets me up in the morning. It gives me purpose. And before I get too old, I hope to someday get back to Ellesmere Island and visit what will soon be the former site of the St. Patrick Bay ice caps. In the grand scheme, these two little ice caps that I became so familiar with back in the early 1980s as Ray Bradley's graduate student are inconsequential victims of the melting Arctic, but the loss is personal. They deserve a decent burial.

NOTES

CHAPTER 1. BEGINNINGS

1. Ray Bradley is an amazing fellow of seemingly limitless energy. He literally wrote the book (a widely acclaimed textbook) on paleoclimatology methods.
2. R. S. Bradley and G. H. Miller (1972), "Recent Climatic Change and Increased Glacierization in the Eastern Canadian Arctic," *Nature* 237: 385–87.
3. R. S. Bradley and J. England (1978), "Recent Climatic Fluctuations of the Canadian High Arctic and Their Significance for Glaciology," *Arctic and Alpine Research* 10: 715–31.
4. J. D. Ives, J. T. Andrews, and R. G. Barry (1975), "Growth and Decay of the Laurentide Ice Sheet and Comparisons with Fenno-Scandinavia," *Die Naturwissenschaften* 62: 118–25. Jack Ives later served on my PhD dissertation committee in 1989; Roger Barry was my academic adviser.
5. J. D. Hayes, J. Imbrie, and N. J. Shackleton (1976), "Variations in the Earth's Orbit: Pacemaker of the Ice Ages," *Science* 19: 1121–32.
6. George was a highly respected scientist, adventurer, and explorer. During his time at the Polar Continental Shelf Program (1972–88), he was immensely proud of the Arctic ice island (a large tabular iceberg), coincidentally named Hobson's Choice, discovered in 1983, which over the next nine years drifted around the Arctic Ocean serving as a platform for scientific exploration. George kept regular contact with Arctic communities, making them aware of scientific projects in their areas of concern, for which he was highly respected. He won numerous awards through the years for his contributions to science, and was also widely known for his musicianship.
7. G. Hattersley-Smith and H. Serson (1973), "Reconnaissance of a Small Ice Cap near St. Patrick Bay, Robeson Channel, Northern Ellesmere Island, Canada," *Journal of Glaciology* 12: 417–21.

8. Fritz Koerner lived life to the fullest. Among his many accomplishments, Fritz was a member of the 1968–69 British Trans-Arctic Expedition, which crossed the Arctic Ocean from Alaska to Spitsbergen via dogsled. He was entirely irreverent, an amazing storyteller, and, from my observations as a field assistant, completely impervious to cold.

9. C. Braun, D. R. Hardy, and R. S. Bradley (2004), "Mass Balance and Area Changes of Four High Arctic Plateau Ice Caps," *Geografiska Annaler* 86A: 43–52.

10. B. C. Forbes, T. Kumpula, N. Meschtyb, et al. (2016), "Sea Ice, Rain-on-snow, and Tundra Reindeer Nomadism in Arctic Russia," *Biological Letters* 12, http://dx.doi.org/10.1098/rsbl.2016.0466.

CHAPTER 2. IT'S NOT WHAT IT USED TO BE

1. NSIDC, "Arctic Sea Ice News & Analysis," https://nsidc.org /arcticseaicenews/.

2. NSIDC, "Greenland Ice Sheet Today," http://nsidc.org/greenland -today/.

3. *The Cryosphere Today*, http://arctic.atmos.uiuc.edu/cryosphere/. They get their sea-ice data from NSIDC.

4. Universität Bremen, *Sea Ice Remote Sensing*, http://www.iup.uni -bremen.de:8084/amsr2/. AMSR2 is also a passive microwave instrument, but it operates at different wavelengths than the passive microwave sensors on the DMPS F-series satellites.

5. Polar Science Center, "PIOMAS Arctic Sea Ice Volume Reanalysis," http://psc.apl.uw.edu/research/projects/arctic-sea-ice-volume -anomaly/.

6. NOAA Arctic Program, *Arctic Report Card*, http://www.arctic.noaa. gov/reportcard/. From the home page, the NOAA Arctic Report Cards can be accessed for all years. Parts of the report cards are also published in the annual American Meteorological Society State of the Climate reports: https://www.ametsoc.org/ams/index.cfm /publications/bulletin-of-the-american-meteorological-society-bams /state-of-the-climate/.

7. The multichannel satellite passive microwave times series is pieced together from three different sources: the Scanning Multichannel Microwave Radiometer (SMMR) aboard Nimbus-7 (October 1978 through July 1987), followed by the Special Sensor Microwave Imager (SSM/I) and the Special Sensor Microwave Imager/Sounder

(SSMIS) aboard the Defense Meteorological Satellite Program F-series satellites. There is also a record from the Nimbu-5 ESMR 5 satellite (Electrically Scanning Microwave Radiometer) from December 1972 through May 1977, but this is based on a single channel, and the sea-ice retrievals are of lower quality.

8. James Maslanik, Chuck Fowler, and Mark Tschudi of the University of Colorado Boulder have spent a lot of effort in putting together this algorithm and constantly improving it.

9. J. E. Kay and A. Gettelman (2009), "Cloud Influence and Response to Seasonal Arctic Sea-ice Loss," *Journal of Geophysical Research* 114: D18204, doi: 10.1029/2009JD011773.

10. M. C. Serreze, A. P. Barrett, and J. Stroeve (2012), "Recent Changes in Tropospheric Water Vapor over the Arctic as Assessed from Radiosondes and Atmospheric Reanalyses," *Journal of Geophysical Research* 117: D10104, doi: 10.1029/2011JD017421.

11. F. Pithan and T. Mauritsen (2014), "Arctic Amplification Dominated by Temperature Feedbacks in Contemporary Climate Models," *Nature Geoscience* 7: 181–85, doi: 10.1038/ngeo2071.

12. International Arctic Buoy Programme, http://iabp.apl.washington .edu/.

13. I. Overeem, R. S. Anderson, C. W. Wobus, et al. (2011), "Sea-ice Loss Enhances Wave Action at the Arctic Coast," *Geophysical Research Letters* 38: L17503, doi:10.1029/2011GL048681.

14. A. Shepherd, E. R. Ivins, G. A. Valentina, et al. (2012), "A Reconciled Estimate of Ice-sheet Mass Balance," *Science* 338, doi: 10.1126 /science.1228102.

15. Back in 2002, Bruce Peterson of the Woods Hole Marine Biological Laboratory was the first to document that discharge from the Arctic-draining rivers in Eurasia was increasing. See the Arctic Report Card: Update for 2015, http://www.arctic.noaa.gov /reportcard/.

16. K. R. Arrigo and G. L. van Dijken (2015), "Continued Increases in Arctic Ocean Primary Production," *Progress in Oceanography* 136: 60–70, doi: 10.1016/j.pocean.2015.05.002.

17. L. S. Guy, S. E. Moore, and P. J. Stabeno (2016), "What Does the Pacific Arctic's New Normal Mean for Marine Life?" *EOS: Transactions of the American Geophysical Union* 97: 14–19.

18. J. C. George, M. L. Druckenmiller, K. L. Laidre, R. Suydam, and B. Person (2015), "Bowhead Whale Body Condition and Links to Summer Sea Ice and Upwelling in the Beaufort Sea," *Progress in Oceanography* 136: 250–62, doi: 10.1016/j.pocean.2015.05.006.

19. NOAA Arctic Program, *Arctic Report Card*, http://www.arctic.noaa .gov/reportcard/.
20. B. C. Forbes, T. Kumpula, N. Meschtyb, et al. (2016), "Sea Ice, Rain-on-snow, and Tundra Reindeer Nomadism in Arctic Russia," *Biological Letters* 12: 20160466, http://dx.doi.org/10.1098 /rsbl.2016.0466.

CHAPTER 3. THE ARCTIC STIRS

1. J. E. Walsh and C. M. Johnson (1978), "An Analysis of Sea-ice Fluctuations, 1953–77," *Journal of Physical Oceanography* 9: 580–91. Following this study, John quickly became a leading Arctic climate scientist, and I have had the honor to work with him on a number of projects. As of the publication of this book, he is still very active. John has remarkable insight and seems to see things that others do not.
2. John Walsh, personal communication.
3. C. L. Parkinson and W. W. Kellogg (1979), "Arctic Sea-ice Decay Simulation for a CO_2–induced Temperature Rise," *Climatic Change* 2: 149–62. The basic thermodynamic treatment of sea ice that Claire and Warren Washington put together as part of the model used in this study is still relevant today.
4. Claire Parkinson, personal communication.
5. S. Manabe and R. Stouffer (1980), "Sensitivity of a Global Climate Model to an Increase in CO_2 Concentration in the Atmosphere," *Journal of Geophysical Research* 85, C10: 5529–54. Suki Manabe was a pioneer in climate modeling, and this paper, now more than 35 years old, is still often referenced.
6. J. Hansen, D. Johnson, A. Lacis, S. Lebedeff, P. Lee, D. Rind, and G. Russell (1981), "Climate Impact of Increasing Atmospheric Carbon Dioxide," *Science* 213: 957–66, doi:10.1126/science .213.4511.957. Some of Hansen's early work was in radiative transfer models, trying to understand the atmosphere of Venus, which has a runaway greenhouse effect. He later refined and applied these models to understand the earth's atmosphere. In his position as a NASA scientist, Hansen became very outspoken over the need to deal with the problem of global warming. Even in retirement, he has been a frequent target of climate change skeptics.

7. A. H. Lachenbruch and B. Vaughn Marshall (1986), "Changing Climate: Geothermal Evidence from Permafrost in the Alaskan Arctic," *Science* 234 (4777): 689–95, doi: 10.1126/science.234.4777.689. This is the first paper I am aware of that provides observational evidence of a warming Arctic.

8. J. E. Hansen and S. Lebedeff (1987), "Global Trends of Measured Surface Air Temperature," *Journal of Geophysical Research* 92: 13345–72, doi:10.1029/JD092iD11p13345.

9. Intergovernmental Panel on Climate Change (1990), *IPCC First Assessment Report*, https://www.ipcc.ch/publications_and_data/publications_ipcc_first_assessment_1990_wg1.shtml.

10. Chris Derksen, personal communication.

11. J. D. Kahl, D. J. Charlevoix, N. A. Zaitseva, et al. (1993), "Absence of Evidence for Greenhouse Warming over the Arctic Ocean in the Past 40 Years," *Nature* 361: 335–37.

12. The first North Pole station (NP-1) was established in 1937, within about 20 km of the North Pole itself. It was led by Ivan Dmitrievich Papanin, polar explorer and scientist, who was awarded the title Hero of the Soviet Union twice and the Order of Lenin nine times. NP stations were then operated continually from 1954 through 1991, when the program shut down with the breakup of the Soviet Union. Some of the drifting NP were located on thick sea ice, while others were on tabular icebergs. From one to three stations were in operation at any one time. The program was re-initiated in 2003. The NP program has been organized by the Russian (former Soviet) Arctic and Antarctic Research Institute.

13. Jim Overland, personal communication.

14. Marika Holland, personal communication.

15. Jen Francis, personal communication.

16. W. L. Chapman and J. E. Walsh (1993), "Recent Variations of Sea Ice and Air Temperature in High Latitudes," *Bulletin of the American Meteorological Society* 74: 33–47, http://dx.doi.org/10.1175/1520-0477(1993)074<0033:RVOSIA>2.0.CO;2.

17. J. D. Kahl, M. C. Serreze, R. Stone, et al. (1993), "Tropospheric Temperature Trends in the Arctic, 1958–1986," *Journal of Geophysical Research* 98: 12825–38.

18. Mike Steele, personal communication.

19. D. Quadfasel, A. Sy, D. Wells, and A. Tunik (1991), "Warming in the Arctic," *Nature* 350: 385, 10.1038/350385a0. The *Rossiya*, from which Quadfasel collected his data, was one of the *Arktika*-class Soviet nuclear-powered icebreakers completed in 1985. A new advanced

class *Arktika* was launched in 2016; the original *Arktika* was retired in 2008.

20. L. G. Anderson, G. Bjork, O. Holby, et al. (1994), "Water Masses and Circulation in the Eurasia Basin: Results from the Oden 91 Expedition," *Journal of Geophysical Research* 99: 3273–83.

21. M. C. Serreze, J. E. Box, and R. G. Barry (1993), "Characteristics of Arctic Synoptic Activity, 1952–1989," *Meteorology and Atmospheric Physics* 1: 147–64. Second author Jason Box was to eventually become a prominent glaciologist. He contributed to this paper when he was still an undergraduate student.

22. J. E. Walsh, W. L. Chapman, and T. Shy (1996), "Recent Decrease of Sea-level Pressure in the Central Arctic," *Journal of Climate* 9: 480–86, http://dx.doi.org/10.1175/1520–0442(1996)009<0480 :RDOSLP>2.0.CO;2.

23. John Walsh, personal communication.

24. J. A. Maslanik, M. C. Serreze, and R. G. Barry (1996), "Recent Decreases in Arctic Summer Ice Cover and Linkages to Atmospheric Circulation Anomalies," *Geophysical Research Letters* 23: 1677–80.

25. The relationship with cyclones seems to be ever more complex. Stormy summers tend to favor retaining ice, but in recent years, it has been clearly established that individual strong storms can sometimes help to reduce ice extent. At least in part, this is probably because the ice is now thinner than it used to be, which has altered its response to winds. Also, strong winds might be able to churn up warmer water from below to enhance melt.

26. T. Haine (2008), "What Did the Viking Discoverers of America Know of the North Atlantic Environment?" *Weather* 63: 60–65, doi:10.1002/wea.150.

27. J. W. Hurrell (1995), "Decadal Trends in the North Atlantic Oscillation: Regional Temperatures and Precipitation," *Science* 269: 6760679.

28. J. W. Hurrell (1996), "Influence of Variations in Extratropical Wintertime Teleconnections on Northern Hemisphere Temperature," *Geophysical Research Letters* 23: 665–68.

29. Jim Hurrell, personal communication.

30. J. T. Houghton, L. G. Meira Filho, B. A. Callander, N. Harris, A. Kattenberg, and K. Maskell, eds. (1996), *Climate Change 1995: The Science of Climate Change*, New York: Cambridge University Press, for the Intergovernmental Panel on Climate Change, https://www .ipcc.ch/ipccreports/sar/wg_I/ipcc_sar_wg_I_full_report.pdf

CHAPTER 4. UNAAMI

1. Don Perovich, personal communication.
2. E. Kalnay, M. Kanamitsu, R. Kistler, et al. (1996), "The NCEP /NCAR 40-year Re-analysis Project," *Bulletin of the American Meteorological Society* 77: 437–71. Atmospheric reanalyses have revolutionized climate science, and the NCEP/NCAR effort was the first such effort.
3. J. Overpeck, K. Hughen, D. Hardy, et al. (1997), "Arctic Environmental Change of the Last Four Centuries," *Science* 278: 1251–56.
4. S. Martin, E. A. Munoz, and R. Drucker (1997), "Recent Observations of a Spring-Summer Surface Warming over the Arctic Ocean," *Geophysical Research Letters* 24: 1259–62.
5. M. C. Serreze, F. Carse, R. G. Barry, et al. (1997), "Icelandic Low Cyclone Activity: Climatological Features, Linkages with the NAO, and Relationships with Recent Changes in the Northern Hemisphere Circulation," *Journal of Climate* 10: 453–64.
6. A. Y. Proshutinsky and M. A. Johnson (1997), "Two Circulation Regimes of the Wind-driven Arctic Ocean," *Journal of Geophysical Research* 102: 12,493–514.
7. M. Steele and T. Boyd (1998), "Retreat of the Cold Halocline Layer in the Arctic Ocean," *Journal of Geophysical Research* 103: 10,419–435.
8. Mike Steele, personal communication.
9. D.W.J. Thompson and J. M. Wallace (1998), "The Arctic Oscillation Signature in the Wintertime Geopotential Height and Temperature Fields," *Geophysical Research Letters* 25: 1297–1300.
10. Jamie Morison, personal communication.
11. "C. L. Parkinson, D. J. Cavalieri, P. Gloersen, H. J. Zwally, and J. C. Comiso (1999), "Arctic Sea Ice Extents, Areas, and Trends, 1978–1996," *Journal of Geophysical Research* 104(C9): 20,837–856, doi: 10.1029/1999JC900082.
12. D. A. Rothrock, Y. Yu, and G. A. Maykut (1999), "Thinning of the Arctic Sea-ice Cover," *Geophysical Research Letters* 26: 3469–72.
13. C. Deser (2000), "On the Teleconnectivity of the 'Arctic Oscillation,'" *Geophysical Research Letters* 27: 779–782.
14. Clara Deser, personal communication.
15. Amanda Lynch, personal communication.
16. M. C. Serreze, J. E. Walsh, F. S. Chapin III, et al. (2000), "Observational Evidence of Recent Change in the Northern High Latitude Environment," *Climatic Change* 46: 159–207.

CHAPTER 5. EPIPHANY

1. Intergovernmental Panel on Climate Change (2001), *IPCC Third Assessment Report*, https://www.ipcc.ch/ipccreports/tar/.
2. I. G. Rigor, J. M. Wallace, and R. L. Colony (2002), "Response of Sea Ice to the Arctic Oscillation," *Journal of Climate* 15: 2648–63.
3. I. G. Rigor and J. M. Wallace (2004), "Variations in the Age of Arctic Sea-ice and Summer Sea-ice Extent," *Geophysical Research Letters* 31: L09401, doi: 10.1029/2004GL019492.
4. W. Krabill, W. Abdalati, E. Frederick, et al. (2000), "Greenland Ice Sheet: High-elevation Thinning and Peripheral Thinning," *Science* 289: 428–30.
5. W. Abdalati and K. Steffen (2001), "Greenland Ice Sheet Melt Extent: 1979–1999," *Journal of Geophysical Research* 106: 33,983–988.
6. J. Zwally, W. Abdalati, T. Herring, et al. (2002), "Surface Melt-induced Acceleration of Greenland Ice-Sheet Flow," *Science* 297: 218–22, doi:10.1126/science.1072708. PMID 12052902.
7. Waleed Abdalati, personal communication.
8. M. B. Dyurgerov and M. F. Meier (1997), "Year-to-year Fluctuation of Global Mass Balance of Small Glaciers and Their Contribution to Sea-level Changes," *Arctic and Alpine Research* 29: 392–402. Mark, who emigrated from Russia and became an American citizen in 2003, was not only a great scientist but one of the most pleasant people to be around whom I have ever met. The community was greatly saddened when he suddenly passed away from a heart attack in 2009.
9. A. A. Arendt, K. A. Echelmeyer, W. D. Harrison, et al. (2002), "Rapid Wastage of Alaska Glaciers and Their Contribution to Rising Sea Level," *Science* 297: 382–85.
10. V. Romanovsky, M. Burgess, S. Smith, et al. (2002), "Permafrost Temperature Records: Indicators of Climate Change," *EOS, Transactions of the American Geophysical Union* 83: 589–94, 10.1029/2002EO000402. Vladimir, another Russian emigre, has been at the University of Alaska since 1992. He is one of the world's leading experts on permafrost.
11. M. Sturm, C. Racine, and K. Tape (2001), "Climatic Change: Increasing Shrub Abundance in the Arctic," *Nature* 411: 546–47. This paper was just a sideline for Matt Sturm. Matt's specialty is snow. Arguably, he knows more about snow than anyone else on the planet. He commonly goes on long transects (many hundreds of miles) across the Arctic via snow machine and sled to

study snow. I am very fortunate to have known Matt for many years.

12. L. Zhou, C. J. Tucker, R. K. Kaufmann, et al. (2001), "Variations of Northern Vegetation Activity Inferred from Satellite Data of Vegetation Index during 1981 to 1999," *Journal of Geophysical Research* 106: 20,069–83.

13. M. M. Holland and C. M. Bitz (2003), "Polar Amplification of Climate Change in Coupled Models," *Climate Dynamics* 21: 221–32.

14. I. V. Polyakov, G. V. Alekseev, R. V. Bekryaev, et al. (2002), "Observationally Based Assessment of Polar Amplification of Global Warming," *Geophysical Research Letters* 29, doi: 10.1029/2001GL011111.

15. O. M. Johannessen, L. Bengtsson, M. W. Miles, et al. (2004), "Arctic Climate Change: Observed and Modelled Temperature and Sea-ice Variability," *Tellus* 56A: 328–41.

16. B. J. Peterson, R. M. Holmes, J. W. McClelland, et al. (2002), "Increasing River Discharge to the Arctic Ocean," *Science* 298: 2171–73.

17. Max Holmes, personal communication.

18. T. J. Boyd, M. Steele, R. D. Meunch, and J. T. Gunn (2002), "Partial Recovery of the Arctic Ocean Halocline," *Geophysical Research Letters* 29: 1657, doi: 10.1029/2001GL014047.

19. K. Y. Vinnikov, A. Robock, R. J. Stouffer, et al. (1999), "Global Warming and Northern Hemisphere Sea-ice Extent," *Science* 286: 1934–37.

20. Don Perovich, personal communication.

21. Jen Francis, personal communication.

22. Charlie Vörösmarty has boundless energy. He has been described as the used car salesman of the hydrologic community because of his ability to sell ideas.

23. Charlie Vörösmarty, personal communication.

24. D. A. Rothrock, J. Zhang, and Y. Yu (2003), "Arctic Ice Thickness Anomaly of the 1990s: A Consistent View from Observations and Models," *Journal of Geophysical Research* 108: 3083, doi:10.1039/2001JC001208.

25. M. Sturm, D. K. Perovich, and M. C. Serreze (2003), "Meltdown in the North," *Scientific American* 289: 42–49.

26. Jim Overland, personal communication.

CHAPTER 6. RUDE AWAKENINGS

1. This has been contended. According to the reviewer of the present book while in manuscript form, NSF had conducted an extensive review of its Arctic program and concluded that it required a major set of budgetary increases that were then implemented.
2. Mike Ledbetter, personal communication.
3. Jamie Morison, personal communication.
4. IASC (International Arctic Science Committee; http://iasc.info) is a non-governmental organization that facilitates international cooperation in Arctic research. IASC is just one of a dizzying and seemingly ever-growing array of U.S. and international organizations aimed at facilitating and coordinating Arctic research and climate research in general. It is a true acronym soup, and it can be hard to determine where one organization ends and the next one starts. Few such organizations have much money—they are largely paper tigers—but they do seem to help, and IASC seems to be one of the more effective ones. IASC was negotiated and established by the foreign ministries of the eight Arctic region countries. However, immediately after signing the founding agreement, the eight-nation foreign ministry signatories signed over IASC to be overseen and managed by the academies of science in those major countries with substantial research programs in the Arctic. It is currently overseen and managed by the academies of 23 nations. The Founding Articles committed IASC, and hence the 23 nations, to pursue a mission of encouraging and facilitating cooperation in all aspects of Arctic research.
5. Mike Ledbetter, personal communication.
6. R. S. Bradley (2011), *Global Warming and Political Intimidation: How Politicians Cracked Down on Scientists as the Earth Heated Up*, Amherst and Boston: University of Massachusetts Press.
7. Raymond Bradley, personal communication.
8. Koni Steffen, personal communication.
9. M. E. Mann, R. S. Bradley, and M. G. Hughes (1998), "Global-scale Temperature Patterns and Climate Forcing over the Past Six Centuries," *Nature* 392: 779–87, doi: 10.1038/33859.
10. M. E. Mann, R. S. Bradley, and M. K. Hughes (1999), "Northern Hemisphere Temperatures during the Past Millennium: Inferences, Uncertainties, and Limitations," *Geophysical Research Letters* 26: 759–62, doi: 10.1029/1999GL900070.
11. Raymond Bradley, personal communication.
12. Charlie Vörösmarty, personal communication.

13. Arctic Climate Impact Assessment – Scientific Report (2005), New York: Cambridge University Press, http://www.acia.uaf.edu/pages /scientific.html.
14. Volker Rachold, personal communication.
15. J. C. Stroeve, M. C. Serreze, F. Fetterer, et al. (2005), "Tracking the Arctic's Shrinking Ice Cover: Another Extreme September Minimum in 2004," *Geophysical Research Letters* 32: L04501, doi:10.1029/2004GL021810.
16. J. T. Overpeck, M. Sturm, J. A. Francis, et al. (2005), "Arctic System on Trajectory to New, Seasonally Ice-free State," *EOS, Transactions of the American Geophysical Union* 86: 309, 312–13.
17. I. V. Polyakov, A. Beszczynska, E. C. Carmack, et al. (2005), "One More Step toward a Warmer Arctic," *Geophysical Research Letters* 32: L17605, 10.1029/2005GL023740.
18. K. Shimada, T. Kamoshida, M. Itoh, et al. (2006), "Pacific Inflow: Influence on Catastrophic Reduction of Sea-ice Cover in the Arctic Ocean," *Geophysical Research Letters* 33: L08605, doi:10.1029/2005GL025624.
19. M. M. Holland, C. M. Bitz, and B. Tremblay (2006), "Future Abrupt Reductions in the Summer Arctic Sea Ice," *Geophysical Research Letters* 33: L23503, doi:10.1029/2006GL028024. Because the climate system is chaotic, very small changes in things such as temperature, wind, and humidity in one place can lead to very different paths for the system. Running a series of simulations with very slightly different starting conditions (an ensemble) provides an avenue to look at the evolution of the climate system as a whole, and, from individual simulations, as well as the spread between all of the simulations, the natural variability in the climate system.
20. J. Stroeve, M. M. Holland, W. Meier, et al. (2007), "Arctic Sea-ice Decline: Faster than Forecast," *Geophysical Research Letters* 34: L09501, doi:10.1029/2007GL029703.
21. Julienne Stroeve, personal communication.
22. Intergovernmental Panel on Climate Change (2007), *IPCC Fourth Assessment Report*, https://www.ipcc.ch/report/ar4/.
23. Recall from chapter 1 that the First IPY, the accomplishments of which were overshadowed by the tragedy of the Greely expedition at Fort Conger, was held during 1882–1883. The second IPY spanned 1932–1933, and the third one, coinciding with the International Geophysical Year, took place between 1957 and 1958. The Fourth IPY was very ambitious and involved thousands

of researchers from 60 different nations. In hindsight, it could have benefited from much better coordination (as one scientist remarked, the approach seemed to be "ready, fire, aim"), but it was a science-driven endeavor, and scientists, as a rule, are not fond of top-down directives.

24. From an NSF fact sheet published in 2007 ("The Arctic Observing Network [AON]," https://www.nsf.gov/news/news_summ .jsp?cntn_id=109687): "The AON is envisioned as a system of atmospheric, land- and ocean-based environmental monitoring capabilities—from ocean buoys to satellites—that will significantly advance our observations of Arctic environmental conditions. Data from the AON will enable the interagency U.S. government initiative—the Study of Environmental Arctic Change (SEARCH)—to get a handle on the wide-ranging series of significant and rapid changes occurring in the Arctic." AON got a big funding boost through the American Recovery and Reinvestment Act of 2009, the stimulus package developed in response to the Great Recession.

25. SEARCH, despite its troubles, was valiantly marching on, providing a focal point for the AON. In turn, a key rationale for AON was the IPY. Jamie Morison's North Pole Environmental Observatory, which had been providing key Arctic Ocean data since its inception in 2000, rolled into the new AON structure, and a bunch of new projects were started. I latched on to an AON project led by Matt Sturm aimed at understanding snow conditions across the North Slope of Alaska. While it was really good science, I was thankful for the involvement as it got me out of the office and into the field for a few weeks each year.

26. J. Wang, J. Zhang, E. Watanabe, et al. (2009), "Is the Dipole Anomaly a Major Driver to Record Lows in Arctic Summer Sea-ice Extent?" *Geophysical Research Letters* 36: L05706, doi:10.1029/2008GL036706.

27. M. Tedesco, M. Serreze, and X. Fettweis (2008), "Identifying the Causes of Greenland's Record Surface Melt in 2007," *The Cryosphere* 2: 159–66.

28. M. S. Rawlins, M. C. Serreze, R. Schroeder, et al. (2009), "Diagnosis of the Record Discharge of Arctic-draining Eurasian Rivers in 2007," *Environmental Research Letters* 4, doi:10.1088/1748–9326/4/4/045011.

29. Julienne Stroeve, personal communication.

30. Dirk Notz, personal communication.

31. R. A. Woodgate, T. Weingartner, and R. Lindsay (2010), "The 2007 Bering Strait Oceanic Heat Flux and Anomalous Arc-

tic Sea-ice Retreat," *Geophysical Research Letters* 37 L01602, doi: 10.1029/2009GL041621.

32. M. C. Serreze and J. A. Francis (2006), "The Arctic Amplification Debate," *Climatic Change*, doi:10.10007/s10584-005-9017.

33. M. C. Serreze, A. P. Barrett, J. C. Stroeve, et al. (2009), "The Emergence of Surface-based Arctic Amplification," *The Cryosphere* 3: 11–19, www.the-cryosphere.net/3/11/2209/.

34. J. A. Screen and I. Simmonds (2010), "Increasing Fall-winter Energy Loss from the Arctic Ocean and Its Role in Arctic Temperature Amplification," *Geophysical Research Letters*, 37, L16707 doi: 10.1029 /2010GL044136.

35. D. S. Kaufman, D. P. Schneider, N. P. McKay, et al. (2009), "Recent Warming Reverses Long-term Arctic Cooling," *Science* 325: 1236–39, doi: 10.1126/science.1173983.

36. M. van den Broeke, J. Bamber, J. Ettema, et al. (2009), "Partitioning Recent Greenland Mass Loss," *Science* 326: 984–86, doi:10.1126 /science.1178176.

37. A. Shepherd, E. R. Ivins, G. A. Valentina, et al. (2012), "A Reconciled Estimate of Ice-sheet Mass Balance," *Science* 338: doi: 10.1126 /science.1228102.

CHAPTER 7. LOOKING AHEAD

1. The provenance of this statement and its variants is actually rather uncertain; it has been ascribed to various individuals, including, and not limited to, Niels Bohr, Yogi Berra, and Sam Goldwyn.

2. Donald Rumsfeld was the U.S. Secretary of Defense under George W. Bush. After his resignation in 2006, he published his autobiography, *Known and Unknown: A Memoir*.

3. E. Kintisch (2016), "Arctic Nations Eye Fishing Ban," *Science* 354: 278.

4. J. C. Stroeve, V. Kattsov, A. Barrett, et al. (2012), "Trends in Arctic Sea-ice Extent from CMIP5, CMIP3, and Observations," *Geophysical Research Letters* 39: L19502, doi:10.1029/2012GL052676.

5. S. Tietche, D. Notz, J. H. Jungclaus, et al. (2011), "Recovery Mechanisms of Arctic Summer Sea Ice," *Geophysical Research Letters* 38: L02707, doi:10.1029/2010GL045698.

6. Dirk Notz, personal communication.

7. M.-L. Kapsch, R. G. Graversen, and M. Tjernstrom (2013), "Springtime Atmospheric Energy Transport and the Control of Arctic Summer Sea–ice Extent," *Nature Climate Change* 3: 744–48, doi: 10.1038/NCLIMATE1884.

8. Q. Ding, A. Schweiger, M. L'Heureux, et al. (2017), "Influence of High-latitude Atmospheric Circulation on Summertime Arctic Ice Loss," *Nature Climate Change* 7: 789-45, doi: 10.1028/NCLIMATE3241.

9. I. V. Polyakov, A. V. Pnyushkov, M. C. Alkire, et al. (2017), "Greater Role for Atlantic Inflows on Sea-ice Loss in the Eurasian Basin of the Arctic Ocean," *Science*, 10.1126/science.aai8204.

10. M. C. Serreze, A. D. Crawford, J. Stroeve, et al. (2016), "Variability, Trends, and Predictability of Seasonal Sea-ice Retreat and Advance in the Chukchi Sea," *Journal of Geophysical Research* 121, doi: 10.1002/2016JC011977.

11. R. A. Woodgate, T. Weingartner, and R. Lindsay (2012), "Observed Increases in Bering Strait Oceanic Fluxes from the Pacific to the Atlantic from 2001 to 2010 and Their Impacts on the Arctic Ocean Water Column," *Geophysical Research Letters* 39: 6, doi:10.1029/2012GL054092. Rebecca is a true ocean-going oceanographer, seemingly always at sea or preparing to go.

12. Marika Holland, personal communication.

13. K. Schaefer, T. Zhang, L. Bruhwiler, et al. (2011), "Amount and Timing of Permafrost Carbon Release in Response to Climate Warming," *Tellus* 63B: 165–80, doi: 10.1111/j.1600–0889.2011.00527.

14. N. Shakhova, I. Semiletov, V. Sergienko, et al. (2015), "The East Siberian Arctic Shelf: Towards Further Assessment of Permafrost-related Methane Fluxes and Role of Sea Ice," *Philosophical Transactions of the Royal Society A*, doi: 10.1098/rsta.2014.0451.

15. Natalia Shakhova, personal communication.

16. The Coriolis force arises because we are on a rotating (hence accelerating) reference frame. Newton's laws of motion hold for inertial (non-accelerating) reference frames. To apply Newton's laws to a rotating frame of reference like the earth's surface (and atmosphere), one must include the Coriolis force. Wikipedia has a rather comprehensive piece on the subject (https://en.wikipedia.org/wiki/Coriolis_force).

17. J. A. Francis, W.-H. Chan, D. J. Leathers, et al. (2009), "Winter Northern Hemisphere Weather Patterns Remember Summer Arctic Sea-Ice Extent," *Geophysical Research Letters* 36: L07503, doi: 10.1029/2009GL037274.

18. J. A. Francis and S. J. Vavrus (2012), "Evidence Linking Arctic Amplification to Extreme Weather in Mid-latitudes," *Geophysical Research Letters* 39: L06801, doi: 10.1029/2012GL051000.
19. E. A. Barnes (2013), "Revisiting the Evidence Linking Arctic Amplification to Extreme Weather in Middle Latitudes," *Geophysical Research Letters* 40, doi: 10.1002/grl.50880.
20. J. Perlwitz, M. Hoerling, and R. Dole (2015), "Arctic Tropospheric Warming: Causes and Linkages to Lower Latitudes," *Journal of Climate* 28: 2154–67, doi: 10.1175/JCLI-D-14-00095.1.
21. S. Lee (2014), "A Theory for Polar Amplification from a General Circulation Perspective," *Asia-Pacific Journal of Atmospheric Sciences* 50: 31–42, doi:10.1007/s13143-014-0024-7.

INDEX

Abdalati, Waleed, 145–147
aerosols, 68, 70; effects of, 109,
139–140, 184; from Mount Pinatubo, 80–82
Agnew, Tom, 79–80
albedo feedback, 6–8, 29–30,
44–46, 68, 76, 83, 113, 183, 213.
See also Arctic amplification
Amundsen, Roald, 15–16
Antarctic Oscillation, 124–125
Arctic amplification, 25, 29–30;
causes and seasonality of,
43–48, 62, 211; emergence of,
138, 152–154, 164–165, 196–199,
203, 232; influences on Greenland ice sheet of, 215; and
mid-latitude weather, 221–227;
model projections of, 151, 206;
and permafrost, 49. *See also* sea
ice extent
Arctic Council, 32–33, 177
Arctic Climate Impact Assessment
(ACIA), 177–179
Arctic Climate System Study
(ARCSS), 78, *See also* Ledbetter,
Mike; National Science Foundation (NSF)
Arctic Dipole Anomaly, 190, 195,
202, 206
Arctic Observing Network (AON),
186–187, 231. *See also* National
Science Foundation (NSF)

Arctic Oscillation (AO), 111–112;
compared to Antarctic Os-
cillation, 124–125; compared
to North Atlantic Oscillation
(NAO), 120–127; mania regarding, 140–142, 156, 163; regres-
sion from high positive state
of, 138, 161–163, 182; and river
discharge, 156, 158; stratosphere
links with, 122–124; sea ice
responses to, 140–142; temperature expressions of, 125–126,
137; unease with, 134–136;
upward trend in, 121–122, 125,
157–158. *See also* climate variability; North Atlantic Oscillation (NAO)
Arctic Transitions in the Land-
Atmosphere System (ATLAS),
132–133, 135. *See also* National
Science Foundation (NSF)
Arrhenius (Svante), 67–68, 160
Atlantic layer and inflow, 45,
88–93; changes in, 90, 93–95,
162, 182–183, 202, 213–214
atmospheric reanalysis, 43,
113–114, 126, 197–198, 222
Azores High, 101–104, 121, 135

Beaufort Gyre, 91, 117, 142, 206
Berra, Yogi, 206
Borenstein, Seth, 233